AF669200

Vertebrate Zoology

Vertebrate Zoology

Dr. B.K.P. Singh

AMIGA PRESS. INC
Delhi -110051

AMIGA PRESS. INC
H.O :- 28-A, Indra Park, Chander Nagar
Delhi :- 110051 (India)
Tel :- 011-22504028, +91-98736-57000
Email :- amigapress@gmail.com

This edition is published by Amiga Press.Inc

ISBN : 978-93-84533-14-4

Copyright© Reserved
First Published 2018

Publisher's note:
This publication may not reproduced, stored in a retrieval system or transmitted in any form or by any means, photocopying , mechanical, electronic, recording or otherwise, without the prior permission from publication house.

Printed and bound in India by Thomson Press Ltd., New Delhi

Preface

Vertebrate zoology is the biological discipline that consists of the study of Vertebrate animals, i.e., animals with a backbone, such as fish, amphibians, reptiles, birds and mammals.

Vertebrate Zoology is the science of animals with backbones, that is, mammals, birds, amphibians, reptiles, and fishes. At the Department of Vertebrate Zoology, we investigate the systematics, phylogenetic history and geographical distribution of those animals. Focal research points include the phylogeny of birds, and the distribution and phylogeny of tropical freshwater fishes.

The subphylum *Vertebrata*includes all of the familiar large animals and some rare and unusual ones as well. The 7 living classesof vertebrates are distinguished mostly on the basis of their skeletal system, general environmental adaptation, and reproductive system.

Three of the vertebrate classes are fish. The most primitive of these is *Agnatha* . It consists of jawless fish that do not have scales. These are the lampreys and hagfish.

Fish that have skeletons consisting of hard rubber-like cartilage rather than bone are members of the class *Chondrichthyes* . These are the sharks and rays. All of the bony fish are members of the class*Osteichthyes* . Tuna, bass, salmon, and trout are examples of *Osteichthyes*.

Animals in the class *Amphibia* spend part of their lives under water and part on land. Frogs, toads, and salamanders are amphibians. Many of these species must keep their skin moist by periodically returning to wet areas. All of them must return to water in order to reproduce because their eggs would dry out otherwise. They start life with gills, like fish, and later develop lungs to breathe air.

The class *Reptilia* includes turtles, snakes, lizards, alligators, and other large reptiles. All of them have lungs to breathe on land and skin that does not need to be kept wet.

They produce an amniote egg which usually has a calcium carbonate rich, leather hard shell that protects the embryo from drying out. This is an advantage over fish and amphibians because the amniote egg can be laid on land where it is usually safer from predators than it would be in lakes, rivers, and oceans.

The class *Aves* includes all the birds. They also produce amniote eggs but usually give them greater protection from predators by laying them high off of the ground or in other relatively inaccessible locations. In the case of both reptiles and birds, the eggs are fertilized within the reproductive tract of females. There are other striking similarities between reptiles and birds in their anatomies and reproductive systems. This is not surprising because birds are descendents of theropod dinosaurs (two-legged mostly carnivorous dinosaurs).

Dogs, cats, bears, humans and most other large animals today are members of the vertebrate class *Mammalia* . All mammals conceive their young within the reproductive tract of the mother and, after birth, nourish them with milk produced by their mammary glands.

Mammals are heterodonts with strong jaws. That is to say, they have a variety of specialized teeth (incisors, canines, premolars, and molars). This allows them to chew their food into small pieces before swallowing it. Subsequently, they can eat any size plant or animal. Many reptiles must swallow their prey whole, which limits them to hunting smaller game.

The book is such that the students will be benefited as far as the their knowledge and examination is concerned.

—Editor

Contents

1

Diversity of Vertebrates

This chapter will be a veritable "parade of taxa", as we start tracing the evolution of vertebrates and the derived characters that distinguish them from the chordates that we discussed in the last lecture.

General Characteristics of Vertebrates

Vertebrates may be characterized by 12 general derived characteristics. You should become very familiar with these traits, and identify how they are expressed in the vertebrates you will see in lab.

1. Bilateral symmetry
2. Two pairs of jointed locomotor appendages, which can include fins (pectoral and anal/dorsal fins, as well as the forelimbs and hindlimbs).
3. Outer covering of protective cellular skin, which can be modified into special structures such as scales, hair and feathers
4. Metamerism found in skeletal, muscular and nervous system. This was described in a previous lecture - structures can include ribs, vertebrae, muscles and ganglia/peripheral nerves.
5. Well-developed coelom, or body cavity completely lined with epithelium (cellular tissue), that may be divided into 2 to 4 compartments.
6. Well-developed internal skeleton of cartilage and bone, separated into axial skeleton (skull, vertebrae, ribs, sternum) and appendicular skeleton (girdles and appendages).
7. Highly developed brain enclosed by skull, and nerve cord enclosed by vertebrae. This provides advanced neural structures that are highly protected from damage.
8. Well-developed sense organs (eyes, ears, nostrils) located on the head (cephalization).
9. Respiratory system, including either gills or lungs, and located closely to the pharynx or throat.

10. Closed circulatory system with ventral heart and median dorsal artery.
11. Genital and excretory systems closely related, utilizing common ducts and pathways.
12. Digestive tracts with two major digestive glands (liver and pancreas) that secrete into it.

Grouping the Vertebrate Classes

Within the vertebrate classes, there are further ways of subdividing the groups based on derived characteristics. There are eight recognized extant classes of vertebrates:

Myxini - hagfishes

Cephalaspidomorpha - lampreys

Chondrichthyes - cartilagenous fishes

Osteichthyes - bony fishes

Amphibia - frogs, toads, salamanders, and caecilians

Reptilia - turtles, snakes, lizards, crocodilians

Aves - birds

Mammalia - mammals

These can be grouped based on their general habitat requirements:

Pisces - collective term for all fishes; includes Myxini, Cephalaspidomorpha, Chondrichthyes, Osteichthyes

Tetrapoda - collective term for the terrestrial vertebrates; they have four feet unless some have been secondarily lost or converted to other uses. Includes Amphibia, Reptilia, Aves, Mammalia

Or based on their feeding habits:

Agnatha - jawless vertebrates, including Myxini and Cephalaspidomorpha

Gnathostomes - vertebrates with jaws derived from the mandibular arch, which may have (in primitive vertebrates) supported gills. Includes Chondrichthyes, Osteichthyes, Amphibia, Reptilia, Aves, Mammalia.

Or Based on their Embryonic Characteristics

Anamniotes - vertebrates that lack an amnion, or extraembryonic membrane that surrounds the embryo and encases it in amniotic fluid. Includes Myxini, Cephalaspidomorpha, Chondrichthyes, Osteichthyes, Amphibia.

Amniotes - vertebrates that possess an amnion. Includes Reptilia, Aves, Mammalia. Don't let these different terms confuse you! They are all ways of distinguishing taxa based on primitive versus derived traits. Use them to help you memorize and classify the taxa we will be talking about.

Tracing Vertebrate Evolution through the Fossil Record

Keep in mind that the evolutionary relationships among the different taxa that we are discussing have been determined from the fossil record. Vertebrates, among all species of animals in the world, have the best fossil record. Reasons include the presence of hard parts, such as bones and scales. The exception to this are the cartilaginous fish, because cartilage does not fossilize well, and the birds, due to the fact that they have hollow bones that can be crushed and lost. In contrast, the best fossil record among the vertebrates exists among the large mammals, whose bones are preserved well as fossils.

Superclass Agnatha

The Agnatha are in some texts referred to as a class, and in others as a superclass. In general this group shares the common characteristics of:

- no jaws
- no paired appendages
- a completely cartilagenous skeleton
- a single nostril
- 6 - 14 external or concealed gill slits
- a persistent notochord
- a two-chambered heart

Because the fossil record is very poor for these species, it is unclear whether the two Agnathan groups should be described as classes or orders.

Ancestral forms of this class were the Ostracoderms, which are extinct, but were heavily-armored on their heads and trunk. The ostracoderms were believed to be detritus feeders, because of their jawless mouths.

Extant Agnathans include two groups, called cyclostomes, because of their circular mouths:

Class Myxini, Order Myxiniformes - hagfishes.

- temperate, marine deep water
- feed on detritus and carrion, as well as polychaete worms
- use sensitive tentacles around their mouths in locating prey
- single nostril opens into pharynx

Class Cephalaspidomorpha, Order Petromyzontiformes - lampreys

- temperate, anadromous (hatch/breed in fresh water, mature in marine) and freshwater
- parasitic as adults - attach to other fishes with their suction-like mouths and rasp a hole in the skin
- buccal glands secrete an anticoagulant to ensure free-flowing food source
- larvae are called ammocoetes, resembling the amphioxus - primarily detritus feeders until they metamorphose into adults, sometimes after 6 or 7 years as a larva.

The remaining vertebrate orders are Gnathostomes (possess true jaws). Evolutionary studies have shown that in most cases the jaw is modified from one of the gill arches that were used to support gills in more primitive species.

Evolution of jaws represents an advancement in morphology, expanding the function of the mouth to a wider range of potential prey types. Thus, the jaws are an example of a derived structure that is more generalized than its ancestral form.

Class Chondrichthyes - cartilagenous fishes characterized by:

- paired nostrils
- skeleton completely cartilagenous with no endoskeletal bone
- no swim bladder
- scales dermal placoid when present
- gill arches internal to gills
- freshwater and marine species

Contains three main groups:

Subclass Elasmobranchii

- 5 - 7 gill openings plus spiracle anterior to first gill
- upper jaw not attached to braincase
- teeth derived from placoid scales, deciduous and continually replaced
- claspers present in males, internal fertilization, ovoviviparous (egg contained within the uterus, where the young develop and then hatch as miniature adults) or viviparous (embryos develop internally and then emerge as a miniature adult)
- modern species present by end of Mesozoic

Order Squaliformes - true sharks

- almost purely predaceous/marine
- heterocercal tailfin - caudal fin is longer on the dorsal side than on the ventral side

Order Rajiformes - Rays, skates, sawfishes

- greatly flattened bottom dwellers
- scales not over entire body
- pectoral fins winglike
- crushing teeth - mollusk eaters
- spiracles greatly enlarged
- oviparous - produce an egg pouch covered in a very tough shell

Subclass Holocephali - Chimaeras

- upper jaw fused to braincase
- flat, bony plates instead of teeth
- operculum covering gillslits
- strictly marine feeding on mollusks

Class Osteichthyes - bony fishes

- endoskeleton made up of bone
- jaws and paired appendages
- gill arches internal to gills
- gills covered by bony operculum
- dermal scales not placoid
- many forms have swim bladder
- appeared in Devonian - dominant vertebrates since mid Devonian
- arose in freshwater, moved into saltwater

Subclass Actinopterygii: ray-finned fish

- fin rays attach directly to girdles
- internal nostrils - nares absent
- single gas bladder
- known from Devonian

Superorder Chondrostei (sturgeons and paddlefish)

- general primitive form
- typically small
- skeleton primarily cartilage
- heterocercal tail
- ganoid scales
- most died out by end of Mesozoic

Superorder Neopterygii (most bony fishes)

Division Ginglymodi - garpike

- moderate ossification of skeleton
- heterocercal tail
- elongated jaws
- ganoid scales
- dominant during Mesozoic

Division Halecostomi

Subdivision Halecomorphi - bowfin

- represented by single freshwater species *Amia calva*
- cycloid scales
- almost homocercal tail

Subdivision Teleostei - true bony fish

- skeleton mostly bony
- tail typically homocercal
- no spiracle
- scales ctenoid or cycloid

- known from marine forms first, originating from Holosteans during Mesozoic
- major radiation associated with modifications in locomotor or feeding mechanisms and high fecundity, i.e.
  - maxilla and premaxilla independently mobile from rest of skull
  - pelvic and pectoral fins adapted for speed and braking giving maneuverability
  - fusiform bodies streamlined for speed

Subclass Sarcopterygii: species previously believed to be extinct, such as the coelacanths and lungfish.

- fleshy lobed fins so that fin rays do not articulate directly to girdles
- internal and external nares
- many retain the heterocercal tail
- the coelacanth is represented by a single species that lives off the Comoro Islands near Madagascar

The rise of the Tetrapoda classes and the movement from water to land represents one of the major evolutionary events in the history of vertebrates. New structural designs were required to make the transition to land in order to cope with increased oxygen levels, decreased water supply, more fluctuating ambient temperature, and slight changes in the way sensory information is obtained.

Class Amphibia - Amphibians

- arose from Crossopterygian, Rhipidistian ancestors
- three extant orders, two extinct subclasses
- lungs and skin used as adult respiratory organs
- gills present in larvae, retained into adulthood in some neotinic forms (salamanders)
- heart with two atria and one ventricle - "three chambered"
- skin is naked or with bony dermal elements
- ectothermic - must regulate body temperature by moving to different microclimates within its environment
- group includes smallest terrestrial vertebrates up to some 5′ in length
- name implies continued tie to water - eggs must be laid in water or at least in very moist environment; young develop as gill breathing, water-dwelling tadpoles
- embryos lack an amnion, but eggs are laid in a jelly-like protective coating

Order Urodela (Caudata) - salamanders

- tail maintained throughout life

- limbs 1 -2 "normal" pairs
- elongated trunk and long tail
- can retain larval characteristics (flattened, shovel-shaped head, fleshy tail, external gills) in adult forms (paedomorphic) - the result is a sexually mature individual with many other body parts in the larval or juvenile condition (neoteny)

Order Salientia (Anura) - frogs and toads

- loose tail as adults
- caudal vertebrae fuse to form long inflexible urostyle - relates to saltatorial locomotion
- long hind limbs developed for saltatorial locomotion
- vocal cords well developed
- ear modified for reception of airborne sound waves

Order Gymnophiona (Apoda) - caecilians

- elongated, snake-like, with no limbs or girdles
- no vocal cords or airborne sound detection
- some retain scales embedded in skin
- notochord persists
- minute eyes, lack lids
- chemosensory tentacle on head

Class Reptilia

- first fully terrestrial vertebrates
- development of cleidoic (closed/self-contained) egg; embryo with extra-embryonic membrane and relatively impermeable shell
- lungs for respiration
- heart with two atria and ventricle partially or totally (Crocodilians) divided
- one occipital condyle
- skin with epidermal scales or bony plates
- ectothermic, sometimes called heliotherms because they can regulate body temperature by using solar radiation
- first appeared in late Paleozoic, so numerous by Mesozoic known as "Age of Reptiles"

Subclass Anapsida ("no opening"); Order Testudinata (turtles)

- ribs modified along with epidermal plates to form shell - carapace and plastron
- girdles inside ribs
- jaws covered with horny epidermal plates, no teeth
- little change since Triassic

Subclass Diapsida ("two openings")

Order Squamata - lizards and snakes

- contains most modern reptiles
- lizards known from Cretaceous, snakes in Cenozoic
- skull has lost one or both temporal regions
- vertebrae usually procoelous
- abdominal ribs usually greatly reduced or absent
- body covered with horny epidermal scales
- quadrate bone moveable
- teeth set in sockets

Subclass Archosauria - Ruling Reptiles

- diapsid skull
- contains dinosaurs and ancestors to birds

Order Crocodilia - crocodiles

- quadrate fixed
- bony plates embedded in epidermis
- teeth set in sockets
- abdominal ribs present in Gastralia
- ventricles completely separated
- developed secondary palate
- "crop" similar to birds

Class Aves

Subclass Neornithes - modern birds

- endothermic rather than ectothermic
- the reptile scale into a feather which is the only unique characteristic of this class
- four-chambered heart
- epidermal scales on bill, legs, feet
- bill instead of teeth; teeth absent in modern forms
- modifications for flight include hollow bones, pectoral appendages modified as wings, air sacs, large eyes and large cerebellum
- modifications for vocalization

Class Mammalia

- possess hair/fur
- mammary glands to nourish young
- endothermic
- viviparous (oviparous in one order)

- two occipital condyles
- zygomatic arch and secondary palate
- single dentary bone in lower jaw
- dentary-squamosal jaw articulation
- muscular diaphragm
- arose from synapsid reptiles which branched off at base of reptilian tree

Subclass Prototheria - egg-laying mammals

- oviparous
- mammary glands without nipples
- cloaca still present
- pectoral girdle with separate precoracoid, coracoid, and interclavicle bones

Subclass Theria

Infraclass Metatheria - marsupial mammals

- viviparous, young born extremely altricial
- abdominal skin pouch (marsupium) supported by epipubic bones
- lack typical chorioallantoic placenta, have yolk-type
- vagina doubled, no cloaca
- mammary glands located inside marsupium
- restricted to New World tropics and Australia

Infraclass Eutheria - "true" or placental mammals

- viviparous
- chorioallantoic placenta
- vagina single
- mammary glands with external nipples
- precoracoid and interclavicle gone
- arose in Cretaceous, great radiation of insectivore-like ancestors during Cenozoic

THE VERTEBRATES

These claim our interest not only as the group of animals with the highest average attainment, not only as the group which contains our own evolutionary pedigree, but because a great deal of the detailed course of their evolution can be traced in the fossil-bearing rocks. All the invertebrate phyla and most of their classes had their origin so far back as to antedate the first fossil-bearing rocks we know.

The earlier stratified rocks, that were laid down when they were first evolving, have been denuded away or so squeezed and baked through heat and pressure that their whole character has been altered, and the fossils they must have contained have been destroyed or rendered unrecognizable. Worms, echinoderms, arthropods, molluscs, corals, lamp-shells, many highly specialized and none essentially unlike those of today, are to be met with in the earliest well-preserved series of rocks, the Cambrian.

But with the vertebrates it is different. Probably their evolution took longer on account of their very complexity; in any case, the first vertebrates so far found belong to a primitive type of fish, and occur in the Ordovician, the next division above (more recent than) the Cambrian. The main features of vertebrate evolution could be deduced equally well from comparative study of present-day forms, or from the anatomy and history of fossils. The two methods confirm each other in all essentials. Starting with fish, the salient steps in vertebrate evolution are as follows:

- The partial conquest of the land by Amphibians, involving the transformation of swim-bladder to lungs, and of paired fins to true limbs with fingers and toes.
- The full conquest of the land by reptiles, no longer restricted to dampness when adult or to water for their early development. This implied the evolution of a large-yolked egg, and the development of a protective water-cushion or amnion over the embryo within the egg. Instead of having to develop in water, each embryo is supplied with what an American writer calls 'its own private pond' in the shape of the fluid within its amnion.

In higher reptiles (mostly now extinct and supplanted by mammals) the body was for the first time raised off the ground and supported entirely by the limbs (or with aid from the tail). Meanwhile, the heart became more or less completely divided into two separate parts, as in man, one for pumping venous and the other for pumping arterial blood only.

- Two separate lines spring from the reptilian stock—the mammals and the birds. Both agree in having developed a mechanism for ensuring a constant temperature-environment for the tissues of the body, and in having the heart completely divided into two. The first to be considered here (though the later to develop in evolutionary time) is the bird line. The evolution of birds was made possible by a series of acquisitions. First, that of constant high temperature, next those of feathers, wings, and air-sacs. In addition, existing birds have lost their teeth, and birdparents show a remarkable degree of care for their young. These steps have led to the chief conquest of the air which has been made by vertebrates.

- The other line led to the mammals. Apart from the constant high temperature and the divided heart, the two universal characters of mammals are the possession of hair, and the secretion of milk by the mother. Further, although a few mammals lay eggs, and a moderate number (most pouched mammals or Marsupials) bring their young up from an extremely early stage in their pouch, stuck firmly on to the nipples, yet the largest and the dominant sub-class of mammals are all characterized by a placenta or organ for ensuring interchange of food, respiratory gases, &c., between the mother and the embryo in the uterus, thus making possible not only a speedier development but also the protection of the embryo within the mother until a later stage than occurs anywhere else in the animal kingdom. All typical mammals are also characterized by having a division of labour among their teeth (incisors, canines, and grinders), which only occurs elsewhere in one extinct group of reptiles.

One or two interesting points emerge. The fishes have continued their unabated success, as the most generally successful group of water-living animals, from very early times up till the present: they hardly compete with mammals or birds. But the Amphibia had their hey-day, and sank with the rise of Reptiles; the Reptiles had a still more marked and more remarkable period of dominance, ending with reptilian collapse and avian and mammalian advance, and the non-human mammals are now showing the same kind of decrease coincident with the rise of man. Thus in the fossil record a succession of types is really visible, and the succession is definitely onc of lower by higher types.

There are, however, other Chordates beside these five classes. Space forbids mention of some of the doubtful 'poor relations' of the stock, and only allows the briefest reference to the Tunicates. These latter are a degenerate group which have lost many of their distinctively Chordate characters. Their degeneration, as in the bivalve molluscs, is due to their having adopted the method of feeding by producing currents and straining off the food-particles; this has led to a sessile mode of life and to loss of sense-organs and diminution of brain.

Many of them form colonies by budding, and have surprising powers of regeneration and dedifferentiation; they are hermaphrodite. They have also lost their skeleton; indeed, each one of them possesses it as a free-swimming larva, and loses it and all the main senseorgans when it metamorphoses and settles down. They must have branched off, however, at a very early period from the main Chordate stem; and it is perhaps comforting to reflect that if they have lost their skeleton, it was by then only a notochord and not a real backbone. A related form, but one much closer to the original primitive

Chordate type, is seen in Amphioxus. The most important things about Amphioxus are those which it does not possess. It has nothing that could be called a head; the merest apology for brain and distance-receptors; no skeleton except a notochord; no heart, but only an ordinary bloodvessel which contracts rhythmically, a liver which is a mere unbranched pocket of the gut, no limbs, no reproductive ducts, the eggs and sperm simply bursting out through the body-wall. It is, however, without a question a Chordate, as shown by its notochord, its pharynx pierced by gill-slits, and its hollow nerve-cord running along the back instead of the belly. It too depends on artificial currents for its food.

It serves to remind us of a time long before that of the earliest fossils preserved to us now, when none of the Chordate stock had reached a higher organization than this; and all the highest types of life—bird, horse, lion, dog, and man himself —were no more than a potentiality slumbering in the germcells of little Amphioxus-like creatures in the sea.

The next stage in vertebrate evolution of which we have any record is represented by the Lampreys; but there is an enormous gap between them and Amphioxus. The Lamprey has already a well-developed brain, a rudimentary skull and backbone of cartilage as well as a notochord, 'nose', eyes, and ears, and a proper vertebrate heart and liver. But they are still far behind any true fish. They have no limbs, no true jaws, no true teeth, and none of them have bone. They, too, like Amphioxus and its relatives, make but a very small group, a relic from the past.

In their life-history, they shed a most interesting light on the evolution of the thyroid gland. The lampern, as the Lamprey's larva is called, still obtains its food from a foodcurrent, in the same way as Amphioxus and the Tunicates. One of the special features of the straining mechanism of all these Chordates is a groove called the endostyle running along the floor of the pharynx, which secretes a sticky mucus.

This is forced forward by cilia, round the mouth in two grooves, and along the dorsal groove back into the intestine. The gill cilia are so arranged that all food-particles strained off by the gills are driven up to the dorsal groove. Here they become entangled and stuck in the slime, and are passed on by this sort of moving stair-case to be digested in the intestine.

The lampern has an endostyle just like that of Amphioxus, except that it is rolled up in a sort of pocket under the pharynx, and sends out its slime-cord ready made.

When the lampern changes into the Lamprey, most of the endostyle degenerates altogether. But some of the cells of its duct remain alive, multiply, and become converted into a typical thyroid. This is a transformation that nobody would have been rash enough to guess at, if they had not been able to see it actually happen, and it is of interest as showing the way in which one organ, no longer required by the animal, may become converted into another organ instead of disappearing altogether.

The balancing planes—the paired fins—of fishes become supporting limbs in land forms; the swim-bladder which regulates a fish's density and consequent distance from the surface becomes a breathing organ—the lung; hair in man has lost its primitive warmth-retaining function, and under the influences of sexual selection has become converted to an adornment. A gap again yawns between the Lamprey and the Fish, although not such a wide one as that between Amphioxus and Lamprey. No fish has a notochord persisting at full size throughout life; all have well-developed vertebrae and skull over-arching the brain, paired limbs, scales, and teeth.

The true jaws have appeared, and can be shown to have arisen by another strange change of function; they are derived from the first pair of the bars of cartilage which support the gills and hold the pharynx-cavity stretched as an open umbrella is held out by its ribs. In almost all fish, however, the upper jaws are not yet firmly united to the skull as in all land forms, but only jointed on. Teeth, too, have an odd history. The skin of dogfishes and sharks can be used for polishing, and when prepared is known as shagreen. Its qualities are due to thousands of little pointed scales sticking out from its surface. When these are examined, each is seen to be nothing else but a miniature tooth fixed by an enlarged base.

Before teeth served as teeth they were scales, and covered the whole surface of the body. It was those in the skin covering the jaws which were able to take on new functions, became true teeth, and eventually alone remained when all traces of the skin-denticles had disappeared.

There are two main groups of fish—the Elasmobranchs (dogfish, sharks, skates, and rays) which have never developed bone in their skeleton, and the Teleosts or higher bony fish, comprising all the familiar species like herring, sole, trout, cod, sea-horse, and flying-fish. The former are primitive in a great many ways. In one respect, however, they are better equipped than the Teleosts.

They lay a few largeyolked eggs, well protected in horny capsules (the 'Mermaids' Purses' one picks up on the seashore) or even in the oviduct of the mother; while the latter lay their eggs before fertilization, and have to produce vast quantities of them in order to compensate for the inevitable wastage during their tiny and unprotected early lives. The cranial nerves are marked with roman numerals. The parts of the brain concerned with various senses are marked as follows: smellcoarse dots; sight crosses; hearing, balance, and lateral line organs-broken oblique lines; touch and other skin senses—vertical lines; taste and other stimuli from viscera (gills, stomach, etc.) horizontal lines.

The motor nerves to gills, stomach, and other viscera are marked in black and white rectangles. It is probable that this one specialization enables the Elasmobranchs to compete not too unsuccessfully with the Teleosts. Their brain is very primitive, even when compared with that of the frog. The smallest fish are under an inch in length, the largest is a form of shark which may reach forty feet.

Fish have evolved into the most extraordinary forms. 'Funny Fish'—pipe-fish, sea-horses, ribbon-fish, sun-fish without a proper tail, cowfish, flying-fish, parrot-fish, porcupine-fish, electric-fish, the flat-fish which fall over on one side as they grow and twist both eyes over on to one side of the head, the remora with a sucker on its head to attach itself to its living locomotive the shark, angler-fish, deep-sea fish with eyes on long movable stalks, or with mouth as big as the whole body, or with rows of red and white phosphorescent lights like a liner at sea.

Fish are the dominant group in the waters, and have become specialized to fill every available niche. One group of freshwater fish, the lung-fishes, are able to survive a long sojourn in the mud when the ponds and streams dry up. This they do by extracting oxygen from the air taken into their swim-bladder, which thus acts as a lung when needed. Thus, these creatures partly bridge the gap to terrestrial life.

In the structure of the limbs, however, a great gap exists between fish and amphibians. No fish has anything but fins, no amphibian anything but legs equipped with fingers and toes. Again, just as the thyroid gland was evolved from the remains of the endostyle when this ceased to be of use, so another ductless gland, the parathyroid, is not found in fish, but only arises out of the debris of the gill-slits when the vertebrates took to land. The Amphibia need not detain us long.

They are in a certain sense a compromise between life in water and life on land. The biggest existing amphibian is the giant salamander, which may reach four feet, though the Amphibia as a group average six inches or less. However, in the Carboniferous period, when they were the only land vertebrates, the average was much higher, and forms over six feet long existed; but all these disappeared as soon as the reptiles entered the field. One other point deserves mention. The Amphibia are the first vertebrates capable of producing vocal sounds deliberately. The reptiles were the true conquerors of the land.

This conquest they owe chiefly to their dry, strong skin, and the evolution of special membranes helping the embryo to live away from water in a largeyolked egg. The allantois, from which the placenta afterwards developed, is the embryo's breathing organ: the amnion is a protective water-cushion, enabling the soft embryo to develop in fluid, protected from pressure and contact.

In their hey-day the reptiles rivalled the present mammals in size and variety of specialization. Besides the existing lizards, snakes, crocodiles, and tortoises, there lived in the middle and late secondary period a whole series of remarkable types. There were mammal-like reptiles, equipped with several different kinds of teeth, and able to run like a typical quadruped; flying pterodactyls; at least two types which had gone back to the sea, the icthyosaurs, which more or less resembled whales (and produced their young alive, as a specimen in the South Kensington Museum testifies, with a brood of embryo skeletons between its ribs), and plesiosaurs, with great flexible necks, creatures which must have looked very much like the average man's idea of the sea-serpent; and C finally, the most successful group of all, the dinosaurs, including rapid runners on two legs, living 'tanks' covered with armour-plate, great semi-aquatic herbivores like Diplodocus , some of which grew to a hundred feet long, and the biggest carnivorous creatures ever known, such as the Tyrannosaurus, which stood over twenty feet high, and no doubt lived upon the gigantic vegetarians. The end of the Secondary period comes, and with the beginning of the Tertiary the pride of the reptiles is humbled. More than half the groups, and those the most advanced, no longer exist; those that are left are already playing second fiddle to the early mammals and birds.

What brought about this revolution is not certain. Possibly an alteration of climate cut down the available food-supply and gave advantages to smaller creatures capable of temperature-regulation. There can at least be no doubt that temperature-regulation and better provision for the young, both before and after birth or hatching, are the two progressive features in which birds and mammals chiefly outdistance their ancestors, the reptiles.

In the case of birds we luckily have come to possess true 'missing links' between reptiles and modern birds. In the mid-secondary, there lived a creature called Archaeopteryx-'Earliest winged creature'. It was an undoubted bird, for it possessed feathers and obvious wings. But its jaws possess a good complement of teeth, the wing is still extremely primitive in possessing claws on three of its fingers (by means of which it no doubt crawled like a bat among the branches), and its tail is not a fan as in all living birds, but is more like that of a kite, with a long jointed skeleton, and feathers coming off on either side all the way down.

Fossil birds found in strata from the close of the Secondary already possessed the modern fan-like tail, and had lost the claws on the wing; but they all still possessed teeth. Toothless birds only appear with the Tertiary period. Flight, high temperature (over 100°, sometimes as much as 105°F), air-sacs and hollow bones, nest-building, bright colours and elaborate courtship, song, and the care of the young—these are the chief characters of modern birds. They owe their success chiefly to one single character—the evolution of feathers. These in the first place keep down radiation and so allow of a high body-temperature. They also permit of the fore-limbs alone being used in flight. Thus the hindlimbs are left free to develop along their own lines, instead of being used up, so to speak, as one of the supports for a wing-membrane, as occurred in the extinct flying lizards, and occurs to-day in the bats. The air-sacs not only lighten the body and help in breathing, but are used to stream-line the body so that it offers least possible resistance to rapid passage through the air. In the same way, most of the fuselage of an aeroplane only serves for stream-lining, not for carrying passengers or goods.

The Pterodactyls had membranous wings like bats, supported both by fore and hind limbs, but mainly by the 'little' finger. The remaining fingers could still be used as claws. Archaeopteryx had true feathers, and wings supported only by the fore-limb, but it retained teeth, a long tailskeleton feathered on either side instead of the tail-fan of modern birds, and three clawed fingers.

Birds are on the average notably smaller than mammals. This is due to purely mechanical aeronautical limitations too complicated to discuss here; as a matter of hard fact, the largest birds capable of flight—swans, vultures, or albatrosses—weigh well under 50 kilograms, while the weight of the great majority is to be reckoned in ounces or even grams, the smallest humming-birds weighing a little under 2 grams.

It is interesting to compare the success of reptiles, birds, and mammals in different zones of the earth's surface. Reptiles have the temperature of their surroundings; consequently their activity is somewhat more than doubled for each rise of 10°C. In the arctic they could scarcely ever be active at all, but would have to exist in a state of almost continuous hibernation, and there are in point of fact no reptiles in the arctic. In the temperate zone they must waste half their life hibernating, and even in the summer cannot compete in activity with a warm-blooded creature; so here reptiles are few and small. In the tropics, however, their average speed of living is more nearly that of a mammal, and they can be active all the year round; in the tropics, therefore, reptiles are more abundant and of greater size—crocodiles, iguanas and other large lizards, giant turtles and tortoises, boa-constrictors and pythons, are all tropical.

Mammals, owing to merely mechanical reasons, can attain to larger sizes than birds. On the other hand, they cannot move readily from one zone to another. So it comes about that in the arctic the birds are the dominant vertebrates, because they can leave in the winter; arctic mammals are few, and almost all of them, like seal, walrus, polar bear, and whale, are entirely or chiefly aquatic. The mammals on the other hand are the dominant group in temperate and subtropical regions.

There are one or two points in connexion with the evolution of mammals that are worth mentioning here. What feathers have been to birds, hair has in part been to mammals; hair and milk together are the mammalian characteristics *par excellence,* hair permitting a constant temperature, milk implying a long period of care of the young after birth. Hair and milk have given mammals the victory over reptiles. But within the mammalian stock itself progress has depended chiefly on two other factors—brain and prenatal care. There are three grades of pre-natal care to be found in the single class of mammals. The duck-bill platypus and echidna lay eggs like any reptile. The Marsupials, such as the kangaroo and opossum, nourish their young within their uterus; but the mechanism is not elaborate enough to permit of its being effective after the embryo has grown to a comparatively small size.

To meet this difficulty, the pouch has been evolved. The embryo is born very small and very unformed (a new-born kangaroo is less than two inches long, naked and blind, the limbs not yet provided with fingers).

It can, however, crawl into the pouch; there it becomes glued to the nipple until it reaches a stage more or less similar to that at which higher mammals are born, then becoming detached but still spending its time in the pouch.

Note the enormous enlargement of the jaws (*pmx, mx,* and *md*) for the attachment of the straining apparatus of 'whalebone'; the fore-limb converted into a paddle (*sc,* shoulder-blade; *h,* humerus; *r,* radius; *u,* ulna); the vestigial hind-limb and girdle embedded in the flank (*p,* pelvis; *f,* femur; the other parts of its skeleton have disappeared); the reduction of the neck (the seven neck-vertebrae, *cu,* are fused into one bone); the fish-like shape, and the tail placed horizontally for rapid diving.

The typical or Placental mammals, on the other hand, while still retaining the milk diet for their young after birth, all possess the wonderful arrangement known as the placenta, by means of which a huge network of blood-vessels formed by the embryo interlocks in the wall of the uterus with a similar network formed by the mother. By this means, although there is no actual passage of blood from mother to embryo, perfect interchange and nutrition is provided, and the embryo can be protected until fully formed.

In some whales the young are retained within the mother's body till they are over twenty feet long. Whales, being aquatic, can attain to much greater size than any terrestrial animal. They include by far the largest animals which have ever existed, at least twice the bulk of the largest extinct reptiles. They also show interesting traces of their origin from land forms in vestigial hind-limbs.

The Placentals have become as dominant over the Marsupials, wherever the two groups have come into contact, as the Mammals over the Reptiles. Only in Australia, which was cut off from the rest of the world by some earth-movement after it had received an invasion of Marsupials but before it had been reached by any Placentals, are the Marsupials dominant—because without placental competitors.

It is very interesting to find that the Australian Marsupials have evolved into a great many forms not found elsewhere, whether fossil or alive, and that the types evolved are often superficially very similar to those of Placental mammals. There is a marsupial wolf, a marsupial mole, the wombat is like a cross between a badger and a bear, some of the phalangers are not unlike squirrels.

The kangaroo, it is true, is of very different construction from any large Placental. It has filled the niche of herbivorous quick-moving animal, but has filled it in a different way from horse or deer. The brains on the left (reconstructed from casts of the interior of the skull) are those of mammals from the early Tertiary period. Those on the right are those of living mammals of about the same total bulk. The two brains of each pair are drawn to the same scale.

Arctocyon (a primitive carnivorous form) A Dog

Phenacodus (a primitive ungulate form) B Pig

Coryphodon (an extinct heavy herbivorous type) C Rhinoceros

Uintatherium (related to Coryphodon) D Hippopotamus

The modern brains, in addition to their increase in absolute size, show an alteration of proportions, the olfactory lobes being relatively rather smaller, the cerebral hemispheres relatively much larger. There are, in fact, the same niches to be filled the world over, and different types may fill them in ways superficially alike or superficially different.

Then there is brain. This has played its chief role in the intense competition which took place in the Placentals. Throughout the Tertiary period new lines of evolution were being developed with great rapidity. The upper limit of size was being increased (as, for instance, in the evolution of the horses, the elephants, the whales, the great cats), and physical specialization, especially of teeth and limbs, was being perfected. Both size and physical specialization, however, soon reached a limit. The supporting power of a limbbone is proportional to its cross-section, i. e. to an area; while the weight to be supported varies as the volume. For purely mechanical reasons, therefore, the limb-bones must become relatively larger with increasing absolute size until they finally grow so unwieldy that size no longer pays. A rhinoceros or an elephant is near the upper margin of size mechanically permissible with advantage to a land animal. As regards specialization, the leg and foot of a horse or a deer, or the teeth of a lion or a cow, could not be much better adapted to their functions than they are now. But, if the instruments at the animal's disposal could not be improved, the methods of using them might be—and this is possible by an improvement in the structure of the brain. The final changes which led to man's evolution seem also to have been primarily brain-changes. Probably, the first divergence of the future human stock from the ordinary land-living mammals came when some shrew-like insect-eating animal took to living in trees.

From some creature like this the Lemur type probably developed, from this again the monkey type. From the old world monkeys, the true apes have clearly descended, by loss of tail and increase of brain-power, and there is no doubt that from some creature which, though not any of the existing apes we know, would have to be classified in the same group with them, man finally evolved. True apes, like the Chimpanzee, are very intelligent and educable. Above, two Insectivores. Note the much greater deve-lopment of the visual region and reduction of the olfactory region in the arboreal Tree-shrew as compared with the terrestrial Jumping Shrew. Centre, the Tarsier, the Lemur nearest to the monkeys; below, one of the most primitive true monkeys. The same tendencies are continued and accentuated. In addition, note the enlargement of the pre-frontal area serving for association.

Taking to the trees appears to have been the necessary preliminary to this long evolution. This put a premium on accurate vision and movements which had to be complicated and accurate if the creature was not to fall and lose its life, while the ordinary land mammals continued utilizing smell more than sight, and turning their limbs into mere supports and running organs.

Only in the trees will there grow up the practice of handling objects carefully, and checking the results by careful examination with the eyes, and this eventually led to the development of a true hand and to the manual skill of human beings. It is interesting to find that the parts of the brain connected with sight and manual dexterity increase in size as we pass from lemurs up through apes to man, while the centres connected with smell decrease very much in relative importance.

But physical acquisitions react upon the mind. The monkey has power of examining an object accurately by touch and sight. As is always the case, it is pleasant to indulge a power that we possess; and hence, it appears, the development of that extraordinary curiosity we all know in monkeys. Their curiosity is largely aimless and useless, but if it could be harnessed to the needs of the race, it might yield the most valuable results, and as a matter of fact, this curiosity was the necessary basis of all man's philosophy and science.

Man himself in all probability developed in some temperate and comparatively treeless region, where the surroundings forced him down out of the easy retreat afforded by the tree tops, and compelled the development of skill, foresight, and reasoning power to cope with the animals that were his enemies and those which, in the absence of fruits, he would have to use for food. The rest of the ape-stock remained in its tropical forest home and was never forced to develop further. Remains of a real link between apes and men, the Pithecanthropus, have been found in Java. In the earlier period of human existence, several species of man, some definitely more simian than any types known to-day, were evolved. But to-day only one species survives.

Man probably originated in the Pliocene. We may mention that prehistoric man is known chiefly by the stone implements which he has left behind; in these a slow but gradually accelerated progress is found with the passage of time. He had to survive the Glacial period, an unfavourable environment which probably served to sharpen his wits; and only about ten thousand years ago at the utmost did he discover the use of metals or the methods of regular agriculture.

After the development of the cell, and the origin of sex (which made variation easier when needed), the first necessary step was the aggregation of cells to form many-celled organisms; without this, neither convenient size nor sufficient division of labour would have been possible. The next steps were precisely those of increasing total size and increasing division of labour among the organs.

First came the establishment of two and then three main layers with different functions, and at the same time the increasing importance of the head end, due to bilaterality and the formation of a nervous system with a dominating region or primitive brain in front. The development of bloodsystem and coelom obviated the need for branched organs, and made much greater size possible. Segmentation again increased the possibilities of division of labour. Increased size made necessary special organs such as heart and gills, while more rapid locomotion was only possible if better sense-organs and better nervous co-ordination was brought about. Ductless glands made possible a new chemical coordination, especially valuable in regulating growth.

Then the emergence from water to land provided a new freedom; temperature-regulation made life stable, and was an absolute necessity for any delicately adjusted mental life. The development of a complicated brain with emotional moods controlling action, made courtship necessary between the sexes; and out of this has developed much of our sense of beauty. The need to develop out of water produced the reptilian egg; and the further prevention of waste of infant life was brought about by the internal development of mammals, permitting the young organism to come into the world at a greater size. As the mechanical efficiency of the organs of the body approached perfection, an increasing premium was put upon more efficient ways of using the organs—in other words upon brain-power.

The most important development in this respect was improved power of learning by experience. But to learn by experience, the youth of the species must be protected and sheltered; hence the extension of parental care to the young for ever longer periods after birth, and the co-operation of both male and female parent in these duties.

Out of this sprang the family, and the constant association on a common task doubtless made the need for communication more urgent, and so was a necessary step towards speech. Then came arboreal life, and the development of dexterity of movement, of the examination of objects by touch and sight, and so of curiosity. Then the re-descent to the ground, with necessity for great self-reliance and skill, and the harnessing of curiosity to be the basis of organized knowledge; with necessity too for more co-operation, and hence of speech, through which alone organized society became possible.

This brief sketch will perhaps give some idea of the strange series of processes, many of them apparently unconnected, which have yet been necessary for human beings to arise, and for mental activity to become the controlling factor in evolution. In conclusion, it should never be forgotten that man is, biologically speaking, quite young. The half-million or million years for which he has been in existence constitute but a small fraction of the time which the non-human mammals, for example, took to reach their highest perfection. Nor is there any reason whatever to suppose that his evolution is over.

However, the one great difference between man and all other animals is that for them evolution must always be a blind force, of which they are quite unconscious; whereas man has, in some measure at least, the possibility of consciously controlling his evolution according to his wishes.

But that is where history, social science, and eugenics begin, and where zoology must leave off.

DEVELOPMENT OF VERTEBRATES

Neurulation

Neurulation is a part of organogenesis in vertebrate embryos. Steps of neurulation include the formation of the dorsal nerve cord, and the eventual formation of the central nervous system. The process begins when the notochord induces the formation of the central nervous system (CNS) by signaling the ectoderm germ layer above it to form the thick and flat neural plate. The neural plate folds in upon itself to form the neural tube, which will later differentiate into the spinal cord and the brain, eventually forming the central nervous system.

Primary Neurulation:

(Shaping, folding, elevation, convergence, closure):

- *Induction*: Response to soluble growth factors secreted by the notochord. Ectodermal cells are induced to form neuroectoderm. Ectoderm sends and receives signals of BMP4 and cells which receive BMP4 signal develop into epidermis. The inhibitory signals chordin, noggin and follistatin are needed to form neural plate. These inhibitory signals are created and emitted by the organiser. Cells which do not receive BMP4 signaling will develop into the anterior neuroectoderm cells of the neural plate. Cells which receive FGF (fibroblast growth factor) in addition to the inhibitory signals form posterior neural plate cells.
- *Shape change/apical constriction*: The cells of the neural plate are signaled to become high-columnar from the surrounding epiblastic ectoderm. The cells move laterally and away from the central axis and change into a truncated pyramid shape. This pyramid shape is achieved through tubulin and actin in the apical portion of the cell which constricts as they move. The variation in cell shapes is partially determined by the location of the nucleus within the cell, causing bulging in areas of the cells forcing the height and shape of the cell to change. This process is known as apical constriction.
- *Folding*: The process of the flat neural plate folding into the cylindrical neural tube. As a result of the cellular shape changes, the neural plate forms the medial hinge point. The expanding epidermis puts pressure on the MHP and causes the neural plate to fold resulting in neural folds and the creation of the neural groove. The neural folds form dorsolateral hinge points (DLHP) and pressure on this hinge causes the neural folds to meet and fuse at the midline. The fusion requires the regulation of cell adhesion molecules. The neural plate switches from E-cadherin expression to N-cadherin expression to recognize each other as the same tissue and close the tube. This change in expression stops the binding of the neural tube to the epidermis.
- *Patterning*: The notochord plays an integral role in the development of the neural tube. Prior to neurulation, during the migration of epiblastic endoderm cells towards the hypoblastic endoderm, the

notochordal process opens into an arch termed the notochordal plate and attaches overlying neuroepithelium of the neural plate.

The notochordal plate then serves as an anchor for the neural plate and pushes the two edges of the plate upwards while keeping the middle section anchored. Some of the notochodral cells become incorporated into the center section neural plate to later form the floor plate of the neural tube. The notochord plate separates and forms the solid notochord.The lateral edges of the neural plate touch in the midline and join together. This continues both anteriorly (towards the head) and posteriorly (towards the tail/caudal). The openings that are formed at the anterior and posterior regions termed neuropores. Failure of the (anterior) and (posterior) neuropore closure results in conditions called anecephaly and spina bifida, respectively. Additionally, failure of the neural tube to close throughout the length of the body results in a condition called cranioarchischisis.

- *Patterning*: After SHH (sonic hedgehod) from the notochord induces its formation, the floor plate of the incipient neural tube also secretes SHH. After closure, the neural tube forms a basal plate or floor plate and a roof plate in response to the combined effects of Shh and factors including BMP4 secreted by the roof plate. The basal plate forms most of the ventral portion of the nervous system, including the motor portion of the spinal cord and brain stem; the alar plate forms the dorsal portions, devoted mostly to sensory processing. The dorsal epidermis expresses BMP4 and BMP7. The roof plate of the neural tube responds to those signals to express more BMP4 and other TGF-b signals to form a dorsal/ventral gradient among the neural tube. The notochord expresses Sonic Hedgehog (Shh). The floor plate responds to Shh by producing its own Shh and forming a gradient. These gradients allows for the differential expression of transcription factors.

Secondary Neurulation

*Neural ectoderm and some cells from the endoderm form the medullary cord. The medullary cord condenses, separates and then forms cavities. These cavities then merge to form a single tube. Secondary Neurulation occurs in the posterior section of most animals but it is better expressed in birds. Tubes from both primary and secondary neurulation eventually connect.

ORIGIN OF THE VERTEBRATES

The story of the origin of vertebrates picks up where the evolution of invertebrates left off. The fact that vertebrates and echinoderm invertebrates both follow a deuterostome pattern of embryo development links the two in the evolutionary framework. From some common echinoderm-like ancestor, all vertebrates are supposed to have risen through the ranks of evolution.

It should be pointed out that evolution is not considered an upward process that ultimately leads to some goal. The bacteria of today could actually be considered more evolved than humans, as they have been evolving for a longer time. Evolution can be described as the information-gaining process that has led bacteria to evolve into birds. Evolution is also described in terms of losing new functions after they originally develop. These apparently contrary forms of evolution, gaining and losing information, highlight the plastic nature of the theory as it bends to accommodate any new evidence. The apparent loss and gain of functions make the evolutionary story contain a multitude of "naturalistic miracles"— if such things can exist. The evolution of dinosaurs into birds is no different from the evolution of flightless ostriches from some flying ancestor. Evolution can be interpreted in any way necessary as long as the underlying assumption that evolution has occurred is not challenged.

In a simplified version of vertebrate evolution, hagfish and lampreys evolved from some invertebrate chordate, the bony fish and cartilaginous fish evolved from some unknown ancestor of the lamprey group, amphibians evolved from an ancestor of the bony fish, reptiles evolved from amphibians, and birds and mammals evolved from reptiles. The details of how each of these steps occurred is an example of historical science. Major assumptions and inferences fill in the gaps of the story. These gaps are commonly referred to as "milestones" or "key adaptations" and are mentioned in the textbooks without any supporting evidence. In the evolutionist's belief system, these milestones must have happened because we can see the results today.

One of the most glaring problems with the evolutionary story of vertebrates is the lack of transitional forms. As in other fossil groups, major phyla appear abruptly and fully formed in the fossil record. The origin of amphibians from bony fish was thought to be demonstrated in the lobe-finned coelacanth fish. Observations of the fish in their deep-ocean habitat has shown that they don't use their fins to walk on the bottom—a trait that allegedly led to their becoming amphibians. Lungfish were also thought to be ancestors, but lung development has made this an unlikely ancestral group. The recent Tiktaalik fossils have provided another option, but the lack of appendicular skeletal support structures (shoulders) and the absence of the hind limbs leaves this an unlikely scenario.

After fish had become amphibians, life was primed to invade the land without relying on the water for reproduction. As reptiles evolved, they needed to adapt to a dry environment to exploit an open niche. Scales were needed to prevent the skin from drying out in the open air and provide protection from harmful UV rays. An efficient lung system had to evolve to allow reptiles to move onto land. The evolution of such soft-tissue structures is one area where inferred stories, based on assumptions, are the only source of information. The absence of preserved soft tissue makes interpreting these structures difficult. Reproduction on land requires the development of some mechanism of internal fertilization. Not only did the eggs need to be fertilized internally, they needed to evolve a structure that would protect them from drying out once they were deposited. The development of this amniotic egg requires the genetic information for many new structures. As has been discussed, natural selection cannot create new information needed for molecules-to-man evolution; it can only select for information (traits) that already exist.

Having accomplished the task of evolving into the reptilian form, birds and mammals come next. Both birds and mammals are supposed to have come from different reptilian ancestors. In the most popular bird evolution hypothesis, theropod dinosaurs (the group to which T. rex belongs) are represented today by all birds. Birds are professed to be living dinosaurs, a distinction formerly reserved for crocodiles. Exactly how this happened in the course of geologic time is disputed at many levels. The media has placed feathers on dinosaurs when none were found with the fossil remains, and every theropod discovered is described as having feathers. The next time you read a story about a dinosaur find and the description includes details of behavior, coloration, and feather patterns, ask yourself what the evidence is for these inferences that are misleadingly presented as facts. In a vast majority of "feathered dinosaur" fossils there is no evidence of feathers. Recent alleged feathered theropod fossils from China have been exposed as frauds or later reinterpreted as not having feathers. The retraction of the original sensational claims is never as obvious as the fantastic headlines that are stored in the memory banks of the public.

Evolutionary models have a difficult time explaining the development of the many unique characteristics of birds from reptiles. The avian lung has a one-way flow that is interconnected to many small air sacs and a countercurrent blood flow system to maximize the efficiency of gas exchange. How this arose from the billows lung of reptiles is a mystery that evolution cannot accurately explain. Feathers are presented in a contradictory manner in the texts reviewed for this book. The Holt text relates that feathers most likely did not evolve from scales; the Glencoe text states that hair and feathers evolved from scales; and the Prentice Hall text implies the evolution of feathers from scales. This reflects the failure of evolutionary thought to accurately describe a key step in vertebrate evolution.

If evolutionary thought provided the explanatory paradigm that its proponents claim, this issue would not be presented in opposite ways in textbooks. Many other issues within the topic of bird evolution—not to mention evolution in general—are disputed among different groups of scientists. While it is necessary for scientists to question each others' statements and independently verify the results, the speculative nature of the evidence and the accompanying explanations leave the debate open to various interpretations. In historical science, like the evolution stories, testing the hypotheses is impossible because they do not represent falsifiable, testable, and repeatable processes. The subjective nature of interpreting supposed evolutionary relationships, whether from cladistics or "molecular clocks," is of questionable validity. Starting with faulty assumptions will lead to faulty conclusions.

In the past, mammals were considered latecomers in evolutionary time. They allegedly hid from the dominant dinosaurs during the day and were only represented by shrew-like creatures that scurried about at night. It was only when the dinosaurs were wiped out that they were able to fill the empty niches. This story has been modified as new discoveries have shown that mammal-like animals (these fossils cannot be evaluated for reproduction and milk production, so they are classified as mammaliaforms) had evolved to live on the land and then invade the water (*Castorocauda lutrasimilis*). This and other discoveries of relatively large mammals have forced evolutionists to rethink the emergence of mammals during the "age of reptiles." Humans are classified as mammals in the evolutionary structure.

Though evolutionists claim to have a clear, big-picture view of the evolution of vertebrates, the devil is in the details. Just because things should have happened in a certain way does not mean that they did. The lack of an information-gaining mechanism is still the most effective argument against the growth of the evolutionary "tree of life." The fossil record and living examples of vertebrates support the creationist "orchard" model, with the origin of all vertebrates on Days 5 and 6 of the Creation Week.

2

Classes of Vertebrates

The subphylum *Vertebrata*includes all of the familiar large animals and some rare and unusual ones as well. The 7 living classesof vertebrates are distinguished mostly on the basis of their skeletal system, general environmental adaptation, and reproductive system.

Three of the vertebrate classes are fish. The most primitive of these is *Agnatha* . It consists of jawless fish that do not have scales. These are the lampreys and hagfish.

Fish that have skeletons consisting of hard rubber-like cartilage rather than bone are members of the class *Chondrichthyes* . These are the sharks and rays. All of the bony fish are members of the class*Osteichthyes* . Tuna, bass, salmon, and trout are examples of *Osteichthyes.*

Animals in the class *Amphibia* spend part of their lives under water and part on land. Frogs, toads, and salamanders are amphibians. Many of these species must keep their skin moist by periodically returning to wet areas. All of them must return to water in order to reproduce because their eggs would dry out otherwise. They start life with gills, like fish, and later develop lungs to breathe air.

The class *Reptilia* includes turtles, snakes, lizards, alligators, and other large reptiles. All of them have lungs to breathe on land and skin that does not need to be kept wet.

They produce an amniote egg which usually has a calcium carbonate rich, leather hard shell that protects the embryo from drying out. This is an advantage over fish and amphibians because the amniote egg can be laid on land where it is usually safer from predators than it would be in lakes, rivers, and oceans.

The class *Aves* includes all the birds. They also produce amniote eggs but usually give them greater protection from predators by laying them high off of the ground or in other relatively inaccessible locations. In the case of both reptiles and birds, the eggs are fertilized within the reproductive tract of females. There are other striking similarities between reptiles and birds in their anatomies and reproductive systems. This is not surprising because birds are descendents of theropod dinosaurs (two-legged mostly carnivorous dinosaurs).

Dogs, cats, bears, humans and most other large animals today are members of the vertebrate class *Mammalia* . All mammals conceive their young within the reproductive tract of the mother and, after birth, nourish them with milk produced by their mammary glands.

Mammals are heterodonts with strong jaws. That is to say, they have a variety of specialized teeth (incisors, canines, premolars, and molars). This allows them to chew their food into small pieces before swallowing it. Subsequently, they can eat any size plant or animal. Many reptiles must swallow their prey whole, which limits them to hunting smaller game.

Like birds, mammals are endothermic , or warm blooded. They are able to maintain a relatively constant body temperature regardless of external environmental conditions mainly by using internal physiological mechanisms.

In other words, they arehomeothermic, or stable in core body temperature, as a result of endothermy. All of the living species of insects, fish, reptiles, and amphibians are ectothermic , or cold blooded. They keep their body temperature in a normal range mainly by avoiding exposure to environmental temperature extremes. For instance, reptiles usually remain in shaded areas on hot days to prevent fatal overheating. On cold nights, their lowered body temperature can cause them to become sluggish and inactive. In contrast, endothermic animals are able to remain active at night and often in the winter when the air temperatures are especially cold.

They can also move about in the heat of very warm days. This ability most likely provided an advantage for the early small mammals in surviving alongside dinosaurs and other large reptiles, which apparently were mostly ectothermic.

The downside of endothermy is the need to consume far more calories relative to body size in order to maintain a constant core body temperature.

Small mammals, such as moles with their rapid metabolism rates, must eat insects or other high calorie foods every half hour or so in order to stay alive. By comparison, cold blooded rattlesnakes usually eat only once every 3-6 weeks and have been known to go without food for as long as two years.

Aiding in mammal body temperature control is their insulating hair and sweat glands. Sweating helps to dissipate heat by evaporative cooling.

Compared to most other land mammals, humans are relatively hairless, but they have far more sweat glands. Mammals have four chambered hearts (like birds), complex nervous systems, and large brains relative to the size of their bodies. This broad range of useful features has made mammals highly adaptive and successful.

They first appeared about 200,000,000 years ago, early in the age of dinosaurs, and replaced reptiles as the dominant class of land animals after 65,000,000 years ago. As the rapidly changing environment at that time led to the mass extinction of most large reptiles, it left vast evolutionary possibilities which mammals took advantage of by rapidly diversifying through adaptive radiation.

Important to mammalian success is their reproductive system. Their bodies took the amniote egg revolution of reptiles and birds one step further. In effect, the uterus functions as the protective eggshell. Young mammals spend a long period of their early development within their mother's uterus. After birth, they are provided with protein and fat rich milk to eat and are usually protected until maturity.

Pregnancy and milk production require mothers to significantly increase their own calorie consumption in order to provide nutrients for their infants. A nursing human female normally uses about 30% of her body's energy just to produce milk.

MAJOR CLASSES OF VERTEBRATES

If you know this much, you know enough to classify any animal you see! Learn these characteristics if you don't already know them. Just think of your favorite example and the characteristics become very easy to remember. The other information gives you more characteristics that scientists use to classify animals into their respective class.

Fish

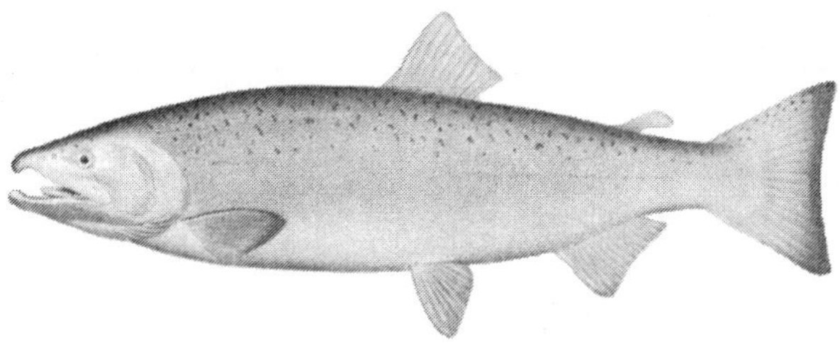

Coho Salmon

1. Fish are ectothermic, aquatic vertebrates.2. Their skin is generally covered with scales.3. Their limbs are modified into fins for swimming.4. They breathe with gills.5. They lay eggs that must be in water.

You probably knew most of this already, didn't you? The term "ectothermic" may be new to you. It means "external temperature" in Latin. The simpler term is "cold-blooded", that is, it has a body temperature equal to the surrounding environment.

You know that fish are cold-blooded if you ever picked one up: they are as cold as the water they were swimming in. Even if you go swimming and are shivering with cold, you are always warmer than the water. "Aquatic" means that they spend their lives in the water.

The most prominent fish at Cazadero is a visitor during the winter rains: the Coho salmon. A federally threatened species, this ocean-dwelling fish comes up Austin Creek when the winter rains raise the water level, to spawn (lay eggs) in the gravel beds that are so typical of the redwood creeks. Note that the eggs are laid in the water, not on land (reptiles and birds lay eggs on dry land).

The eggs drift down into the gravel, where they are protected. The eggs at the left are Coho salmon eggs, and they show the developing embryo, and you can see the eye of the baby fish.

A salmon is a fish because it is ectothermic (cold-blooded), lives in the water, breathes with gills, lays eggs in water, and has fins and scales.

Amphibians

Pacific Treefrog

1. Amphibians are ectothermic vertebrates.
2. Their skin lacks scales, hair, and feathers, and is either smooth (like a frog) or rough (like a toad). They are dependent upon moisture and subject to desiccation; their skin must remain moist to aid in breathing.
3. They lay eggs in water, which hatch into an intermediate life form (tadpole or larva) that usually breathes with gills, and change into the adult form that breathes air and can live outside water.
4. They have three-chambered hearts.
5. They lack claws on their toes.

You already knew all the facts that are in bold, didn't you? Let's think of the various kinds of amphibians that Cazadero has, and you can probably figure out all the characteristics of the class. The frog at the top is the Pacific treefrog, one of the most famous frogs in the world. Do you know why? It is the frog heard round the world! Most Hollywood movies have used the croaking of Pacific treefrogs when they want frog noises, because they are champion croakers. They are treefrogs, so they do not spend much of their time in water, but are found in trees.The toad (below, left) is the Western Toad, which you can find in the thousands down by Austin Creek in later summer when they have hatched out from their tadpole stage. The amphibian (below, right) is the California Newt. Newts are like salamanders, but they have the rougher, dryer skin like a toad. Frogs are like salamanders, and toads are like newts. They all have moist skins, through which they can breathe (they also have lungs for breathing).The salamander at the bottom is the California Slender Salamander, which is about 3 inches long but only 1/4 inch wide.The name "amphibian" comes from the Latin for "two lives", which refers to the fact that all amphibians have a larval stage (we call the larval stage of frogs and toads a "tadpole" or "pollywog") and an adult stage. This is unique in the vertebrate world, because fish, reptiles, birds, and mammals are all born as small versions of the adult, not a completely different form. This dual life, one aquatic with gills and one terrestrial with lungs, may be representative of the evolution of animals out of water and onto dry land.Now, lets look at the characteristics of the class amphibian. You already knew that frogs were cold-blooded if you ever picked one up. You also knew that tadpoles hatched into frogs, and that frogs do not have feathers, scales, or hair ("fine as frog fur" is a famous phrase), and that their skin was moist. You probably already knew that frogs lay eggs. So you are already an expert on what makes an amphibian an amphibian! The other characteristics may be fun to know, but you can classify an animal as an amphibian just by what you already know.

Frogs, toads, salamanders, and newts are all amphibians because they are cold-blooded, have no scales, hair, or feathers, their skin is moist, they lay eggs in water, and their life cycle has two stages, the "pollywog" or larval stage and the adult stage.

Reptiles

Garter Snake

1. Reptiles are ectothermic vertebrates.
2. Their skin has scales, but no hair or feathers.
3. They have three-chambered hearts (except for alligators and crocodiles, which have four-chambered hearts).
4. They have claws on their toes (except those which do not have legs, such as legless lizards).
5. They are the first animals, in evolution, to develop the amniotic egg. This allows reptiles to lay eggs on land.

You already knew all the facts that are in bold, didn't you? Let's think of the various kinds of reptiles at Cazadero, and review what we already know about them.

The most common reptile that you can find at Cazadero is the alligator lizard, usually crawling though the leaves and debris on the ground, looking for invertebrates to munch and lunch on. You may also see garter snakes along the creek side, blue belly lizards sunning themselves on rocks or logs, or the rare turtle in the creek.What do all these reptiles have in common? They all are cold-blooded (the reason lizards and snakes sun themselves is to warm up, just like you on a cold morning, or after a wade in Austin Creek) and they have scales. Did you ever see a lizard with feathers, or a hairy snake? No! Scales make up their only covering. Reptiles also lay eggs on dry land. You have heard of turtle eggs, right? Nature shows often have programs on sea turtles and how they lay their eggs on the beach. You may have seen some nature programs on alligators, another kind of reptile but not one we have, laying eggs in piles of rotting vegetation. All reptiles lay eggs. We know now that birds evolved from reptiles, and one of their linkages is the fact that they both lay hard-shelled eggs on dry land. Amphibian eggs, in comparison, are soft and must be laid in water.

Snakes, turtles, and lizards are all reptiles because they are cold-blooded, they lay eggs on dry land, and are covered with scales, never feathers or fur.

Birds

You already knew all the facts that are in bold, didn't you? Everyone knows birds have wings with feathers and lay eggs. You may not have known they are endothermic, which is Latin for "inside temperature". This means that they can maintain their own warm inside temperature, even when it is cold outside. Frogs (amphibians) and snakes (reptiles) have to hide when the weather is cold, but birds can be out and active. Let's think of the various kinds of birds at Cazadero, and think what we already know about them. The most common birds at Cazadero are blue jays, both Scrub Jays and Steller's Jays, robins (at right), as well as a variety of aquatic birds such as ducks, mergansers, and herons, which feed along Austin Creek.The characteristics of the class

Birds are rather well-known and hardly need to be taught. We all know that birds have wings, chickens lay eggs, and that birds of a feather flock together. The only character that you might not have known is that they are warm-blooded, unless you or a friend have a pet bird at home or if you have held a chicken. Birds in general have a higher body temperature than we do, so they feel quite warm when we hold them. Their feathers are well-adapted to hold this heat in, which is why we make comforters out of down.

Mammals

Cazadero has a lot of mammals, but they tend to live elsewhere and only move through the Redwood Forest. There is not much in the way of natural food in the redwoods, but camp provides a lot of human food, so raccoons and mice are frequent guests. The opossum and deer might wander along Austin Creek.

Deer, mice, opossums, raccoons, and humans are mammals because they are warm-blooded, have hair, and nourish their young on milk.

CLASSIFICATION OF VERTEBRATES

Biology - Animal Classification

The earth is populated with a tremendous variety of animal life. In a process called classification, scientists organize the incredible diversity of animal life into a manageable body of knowledge that allows for systematic study. When scientists classify animals, they sort them into groups, on the basis of similarities and differences in physical structure, biochemistry, genetic composition, and other factors.In this program we will discuss a system in which animals are classified into seven groups, from kingdom to species.

As one goes from kingdom to species, the animals in a group are more closely related. For instance, although these animals appear to be radically different from each other, they share certain physical and structural characteristics. These similarities have prompted scientists to classify all three into the phylum Chordata, which contains fish, amphibians, reptiles, birds, and mammals, generally considered by scientists to be complex.One of the primary characteristics of the members of this group of animals known as chordates is the existence of an internal supporting rod of connective tissue.

This rod of connective tissue is called a notochord. In most chordates, the notochord is only present during the earliest, or embryonic, stage of development, as seen here in this frog tadpole.A second major characteristic of all chordates is the existence of a tubular nerve cord that runs along the top, or dorsal side of the body, as, for example, the spinal column that runs along the back of each human being.A third important characteristic of chordates is paired gill slits, which often as in the case of humans, exist only in the embryonic stage of development before changing into human structures. However, certain animals, such as fish, never develop an organ for breathing beyond gill slits.

The members of the phylum Chordata with which we are most familiar are those further classified into the subphylum Vertebrata. In all vertebrates the notochord is replaced during the embryonic stage by a column of bone or cartilage, the backbone. Vertebrates also possess a skull that protects the brain. Vertebrate skulls house the sensory organs for taste, smell, hearing, and vision. Most vertebrates have two pairs of appendages which can take the form of wings, flippers, fins, arms, or legs.Externally, vertebrates are covered with a double layer of tissue. The external layer, or epidermis, can produce scales, as is often the case in reptiles and fish; feathers, which are a common feature of birds; or hair, which is found on a variety of animals like seals and horses, not to mention humans.

Amphibians

Another group of vertebrates are the members of the class Amphibia.

The word Amphibia means "double lives." They live both on land and in the water. Amphibians are cold-blooded animals with a three-chambered heart as well as other organs commonly found in most vertebrates. Typical examples of amphibians are frogs, toads, and salamanders. Most amphibians have soft, smooth skin, which is kept moist by mucous glands.

Some, however, such as the toad, have dry, bumpy skin.Amphibians absorb oxygen into their bodies through their skin. Amphibians also have lungs, which are used in respiration, but they are not as highly developed as those of the higher vertebrates. Amphibians usually live in damp places, which helps to keep their skin from drying out. In order to mate successfully, amphibians must be in a watery environment. Most amphibians are oviparous.

They develop from eggs hatched outside them other. Amphibian eggs can be found in the spring and early summer as clumps of jelly-like masses laid in freshwater. Amphibian eggs, such as the frog egg shown here, undergo many cell divisions. In time they will develop into fishlike larvae called tadpoles. In their younger stages, tadpoles use gills for respiration.As an amphibian's limbs begin to develop and the tail disappears, the respiratory function of the gills is replaced by the lungs. Still, some salamanders retain external gills as adults.

Upon the completion of its metamorphosis, a tadpole is transformed into an adult frog, capable of life not only in water but also on land. This capability is a step of great evolutionary importance. The tadpole's chief food is vegetable matter. Upon becoming frogs, they most often feed on animals such as these dragonflies, which they capture with their long, sticky tongues.Amphibians protect themselves by quickly scurrying away from danger. Some amphibians also have foul-tasting or poisonous skin that serves to deter predators. In terms of evolution, the classes containing fish and amphibians are some of the simplest vertebrates in the phylum Chordata. And yet these animals exist in a wide array of beautiful shapes and colours, and comprise an essential part of the network of all living things.

Other members of the subphylum Vertebrata include birds in the class Aves, reptiles in the class Reptilia, and mammals in the class Mammalia. These animals breathe exclusively through the use of well-developed lungs, rather than through the use of gill slits, as do fish, or through their skin, as do amphibians.

Birds

Looking at the extinct Archaeopteryx, which exists today only in fossil form, we can perhaps see the way certain physical characteristics from each group have been merged. Feathers, which come in all shapes, sizes, and colours, are certainly an important distinguishing characteristic of birds. Feathers serve many useful purposes, including insulation for the body and protection from moisture. Their special coloration may be useful for concealment or for attracting a mate.

Like mammals, birds are warm-blooded. The temperature of a bird's body remains the same regardless of the surrounding air temperature. Birds must consume a great deal of food to maintain their body temperature, which is slightly higher than that possessed by mammals. Birds have a highly efficient respiratory system and a fast beating, four-chambered heart.Most birds are capable of flight, although a few, such as the ostrich and penguin, do not now have this ability. It is interesting to examine the adaptation that birds have undergone in order to fly.

As their forelimbs have developed into wings, large chest muscles have developed which originate at the keel of the breastbone. Many of a bird's bones are hollow and lightweight. Their bodies are also streamlined to avoid wind drag while in flight. The tail has also been modified for flight and is used for steering, braking, balancing, and lift-off.In addition to flying, birds have adapted in other ways to serve other purposes.

For instance, rear legs are used for walking, and the feet have adapted to serve purposes such as paddling, perching, wading, and tearing the flesh of prey. Because birds have no teeth, their beaks have undergone certain modifications required for survival. Some beaks are used for ripping flesh, some for cracking seeds, and some for boring in wood or capturing fish. Birds can communicate through a variety of songs and calls that are produced through a unique organ in their throat called a syrinx.Birds have other unique characteristics as well.

Birds are oviparous and lay eggs with brittle shells which must be incubated while they develop. Most young birds require parental care until they are old enough to fly and get their own food. Birds have keen eyesight, which is helpful in protection and in obtaining food, such as small animals and insects. Some birds, such as vultures, are scavengers, meaning they prey upon dead or decaying matter.

Fish

Fish and amphibians are vertebrates with special characteristics that allow them to spend all or part of their lives in water. There are two classes of fish in the subphylum Vertebrata.The first of these classes contains sharks, rays, and skates and is called class Chondrichthyes, the cartilage fish. Members of this class do not have bone in their skeletons. Instead, they have skeletons of rubbery cartilage.The second class of fish are those with bony skeletons. These are members of the class Osteichthyes, and include most of the fish familiar to you, such as salmon and trout.Many fish have a body shape that easily facilitates movement through the water. The tail is a chief "organ" of locomotion, while the fins are used to steer, swim, and maintain balance. Overlapping scales cover the bodies of most fish, although some fish, like sharks, have a rough leathery skin. Special organs called gills enable fish to obtain oxygen dissolved in the water. In most fish the gills are protected by covers called opercula. When water enters a fish's mouth, it passes over the gills where tiny blood vessels absorb oxygen from the water and release carbon dioxide into it. Water is then expelled through the gill slits.The internal organs of fish are more highly developed than those of invertebrates, animals without a notochord, dorsal spinal cord, or gill slits. For instance, fish have a two-chambered heart, a liver, and, in many cases, an air bladder, which is used to regulate depth. Another characteristic of most fish is that their body temperature is similar to the water they live in; thus we call fish cold-blooded animals. Most fish are oviparous, or egg-laying. The female lays eggs in the water where they are fertilized and left to hatch. Certain sharks are ovoviviparous. This means their eggs remain within the mother's body until they hatch into baby fish, after which they are "born." Fish eat plant material as well as other aquatic animals. They even eat one another.

Mammals

Along with the fish, amphibians, reptiles, and birds we've already looked at, mammals are also classified as vertebrate members of the phylum Chordata. The giraffe is one of a variety of species that belong to the phylum Chordata, class Mammalia. Because of their refined nervous and reproductive systems, mammals are generally considered the most advanced members of the animal kingdom. After they are born, mammals are nourished with milk secreted by the mammary glands of their mothers. It is for this unique characteristic that mammals are named.A second characteristic of mammals is that they possess hair that can cover all or part of their bodies. Most mammals are viviparous. This means that mammalian young develop inside the mother's body rather than outside, such as in an egg. Mammalian offspring are born in a form resembling an adult.Some subclasses of mammals give birth to their offspring by other means. The rarest are the monotremes, or egg-laying mammals, which includes the platypus and the species known as echidna, or spiny anteater. Another subclass includes the marsupials or pouched mammals such as the kangaroo and opossum. Marsupial offspring are born in an undeveloped state and spend the last period of development inside the mother's pouch, where the mammary glands are located.All members of the class Mammalia are warm-blooded and are able to maintain a fairly constant body temperature. All mammals are fundamentally alike in body structure. Special characteristics include a well-developed brain, a four-chambered heart, and a muscular diaphragm that separates the abdominal and thoracic cavities. Most mammals have two sets of limbs, which have adapted for special needs such as grasping, flying, swimming, and walking. Many mammals grow two sets of teeth.

The first set, known as milk teeth, are actually pushed out by the permanent teeth.Mammals come in all sizes, from tiny rodents to enormous elephants. Mammals also live in a variety of habitats with bodies that are adapted to meet many environmental conditions. Some aquatic mammals, such as the seal and the sea otter, have adapted to life in salt water, while others, such as beavers, make their homes in freshwater. Some mammals live in treetops where it is useful to possess grasping hands, feet, and tails. Still other mammals, such as prairie dogs, live in small villages underground.Within the category of mammals, those that possess hooves make up a very large and diverse group. Hoofed mammals have teeth with broad surfaces that are used for grinding substances such as grass and other leaves. Hoofed mammals are classified into two orders according to the number of toes they possess. Animals in the order Perissodactyla have an uneven number of toes. Members of the order Artiodactyla, such as camels, possess an even number of toes. Although camels thrive in dry, hot environments, there are some mammals, such as polar bears and other members of the meat-eating order Carnivora, which flourish in colder climates. Other members of the order Carnivora include cats, both tame and wild.The largest land animal in existence is the elephant, which is classified within the class Mammalia, order Proboscidea, because of its long and unusual nose. A widely recognized group of mammals are those within the order Primates. Monkeys, apes, and man are included in this category of animals. All primates are able to walk in an upright position. Among the larger mammals, Homo sapiens are undoubtedly the most successful and unique. The human population of the earth already exceeds 5 billion.

Reptiles

Members of the class Reptilia are regarded by scientists as more sophisticated than either fish or amphibians, but less sophisticated than either birds or mammals. Some of the more common members of the class Reptilia are turtles, snakes, and lizards.A major characteristic of reptiles is that they have either an outer covering of scales or thick, leathery plates. All reptiles, whether they live on land or in water, breathe by means of well-developed lungs.

Reptiles such as turtles, alligators, and certain snakes spend most of their lives in water. But even aquatic reptiles must come to the surface for an occasional breath of air. All modern reptiles are cold-blooded. Therefore, their body temperature changes with the temperature of the environment in which they live. Reptiles are sluggish in cold weather. They like to warm their bodies in the sunshine. In very hot climates, reptiles seek shade to keep from overheating. During cold winters, reptiles become inactive and hibernate.

They emerge from their burrows only after the earth warms in springtime.Since the time of the dinosaurs, the great majority of reptile species have become extinct. Only four of the nine original orders survive. These are lizards and snakes, turtles, crocodilians, and tuatara. Tuatara are lizard like reptiles that are native to New Zealand.Among the reptiles that survive today, most are oviparous, which means that their young develop from eggs which are laid outside the mother's body. However, there are also certain snakes which are ovoviviparous, meaning that the mother's eggs hatch inside her body. These offspring are delivered alive. A few snakes are poisonous. Some, like rattlesnakes, have an unusual organ in their heads for detecting warmth. This gives them a special sensing ability when searching for warm-blooded animals, such as rodents. For this reason snakes are beneficial to farmers whose grain crops can be seriously damaged by rodents.The largest reptiles of all, the dinosaurs, became extinct a very long time ago. Dinosaurs first appeared on the earth over 200 million years ago. They flourished during what is now known as the Age of Reptiles. Like the lizards of today, some dinosaurs, such as the Tyrannosaurus rex, survived by eating other animals, while some, like the Brachiosaurus, survived on plant life. There were also fishlike dinosaurs, and dinosaurs capable of flight. Numerous experts now believe that dinosaurs were probably warm-blooded animals like modern birds and mammals. However, they are still placed in the class Reptilia. Humans have developed an intelligence that enables us to use the raw materials of the planet to suit our own unique needs. We can transform rock into metals and glass, and use the metals to construct various forms of transportation. We can fly to the moon, create brilliant art, and provide food and nourishment for our citizens.

But at the same time, humans have also created tremendous problems, from war to air and water pollution. Humankind can destroy as well as create. It is up to present and future generations to make sure that the earth is kept in proper working order, so that life may continue to grow, develop, and flourish on our planet.

GLANDS AND SCALES OF VERTEBRATES

Introduction

In all vertebrates the skin has two major layers. The outer, relatively thin epidermis is composed of closely packed cells with little intercellular material; it provides the barrier against attack by chemicals, radiation, or microbes.

The underlying dermis (cutis, corium) is thicker and tougher, and its bulk is formed by extracellular materials manufactured by scattered cells. One of its major functions is physical protection.

The sensory functions of skin are shared by both epidermis and dermis. The various structures or appendages such as scales, feathers, claws, glands, and hair follicles are derived from either layer or from both. There are considerable differences between the skins of different vertebrate classes, and they are closely related to the environments of the various groups.

The epidermis is the product of the deepest layer of its cells, those that lie immediately over the dermis. From this generative layer, known as the stratum germinativum, cells move outward and become progressively flattened. The surface cells of terrestrial vertebrates, mere remnants of once living cells, are scaly and compressed; they constitute the horny layer, or stratum corneum. The cell fragments of the stratum corneum are composed largely of keratin, a tough insoluble protein. In most land vertebrates the stratum corneum is shed or molted, either periodically and in large fragments or sheets, as in reptiles, or continuously in small patches or scales, as in mammals.

The dermis, which is best developed in mammals, consists largely of fibrous connective tissue (composed of collagen fibres), blood and lymph vessels, smooth muscle cells, and nerve endings. It gives rise to so-called membrane bones—the bony scales of fishes, the bony plates in certain reptiles and mammals, and the membrane bones of the vertebrate skull. Through its blood network the dermis supplies nourishment to the overlying epidermis.

Skin Derivatives and Appendages

Among the notable changes that have taken place during the course of evolution is the development in vertebrates of a variety of glands, pigmentary structures, scales, claws, nails, horns, feathers, and hairs as adaptations to their changing environments.

Skin Glands

The glands of the skin are all exocrine, that is, they secrete their products, usually through ducts, to the epidermal surface. They may be unicellular, as are the goblet cells of fishes, or multicellular, as are the sweat glands of humans. Some multicellular glands are tubular and extrude their secretion into a central space or lumen; some, like the oil-producing sebaceous glands of mammals, form their product by complete breakdown of the cells, a method of secretion known as holocrine. Glands may consist of tubes or sacs, and they may be singular, clustered, or branched; some even contain units of more than one type. They may secrete their product continuously, periodically, or only once. Mucous glands secrete a protein called mucin, which with water forms the substance known as mucus; this slimy material serves to lubricate the body, thus lessening friction and aiding locomotion in swimming animals. Serous glands produce a watery secretion; sweat glands of mammals are of this type. Sebaceous glands secrete oil, ceruminous glands secrete wax, mammary glands secrete milk, poison glands secrete various toxins, and scent glands secrete a variety of odoriferous substances. Further, certain epidermal glands may be modified into light-producing structures called photophores, seen in the skin of many deep-sea fishes.

Pigment Cells

In fishes, pigment is produced in branched cells known as chromatophores, which can be found in both epidermis and dermis. Rapid colour change, by which some fishes can adapt to a change of background, is brought about by redistribution of the pigment within the cell boundaries. Slow, long-term changes involve alterations in the numbers of cells or in the amount of pigment they contain.

Chromatophores are also present in amphibians and reptiles, but not in birds or mammals, which possess pigment cells called melanocytes. Melanocytes are found mainly in the epidermis, though they occur elsewhere.

They also are branched, or dendritic, and their dendrites are used to transfer pigment granules to adjacent epidermal cells. A number of different pigments are produced in the different vertebrate groups, but in mammals only brown eumelanin and yellow or red phaeomelanin are important. Pigment cells—chromatophores and melanocytes alike—are influenced by melanocyte-stimulating hormones of the pituitary.

Epidermal Scales

Epidermal scales are horny, tough extensions of the stratum corneum. Well developed in reptiles, they are also common on exposed skin in birds and mammals. Such scales are periodically molted or shed gradually along with the rest of the stratum corneum. Epidermal scales are absent in fishes, but dermal, or bony, scales are abundant. Clawlike epidermal scales are present in certain amphibians, including a few toads, certain burrowing, wormlike caecilians, and the salamander Hynobius.

The so-called horns of the horned lizard are specialized epidermal scales; and the rattle of rattlesnakes is a series of dried scales loosely attached to each other, the last one always remaining despite molting of the rest of the stratum corneum. Epidermal scales cover the bony scales of the carapace and plastron of turtles' shells. The beak of turtles is composed of a modified epidermal scale covering the jawbone.

Bills of birds are similarly constructed. In birds, epidermal scales are confined to the lower legs, feet, and base of the bill. The spurs of some birds are bony projections covered with a scalelike sheath. The skin of the webs in aquatic birds is also scaly. In mammals, except for a few cases, epidermal scales are largely restricted to the tails and paws. The overlapping horny plates of the pangolin are modified epidermal scales.

HORNS AND CLAWS OF VERTEBRATES

Introduction

In zoology, a horn is one of a pair of hard, pointed, often permanent projections on the head of various hoofed mammals (ungulates) consisting of a core of living bone covered by a sheath of keratin and other proteins. Examples of animals with true horns include antelopes, cattle, buffalo, and goats. The term also is used to refer more specifically to the hard keratinous material forming the outer covering.

While both a bony core and a covering of keratinous material constitute the definition of a "true horn," there are many other hard structures projecting from the head of animals that do not meet these requirements and yet are referred to commonly as horns. These include the antlers of deer (dead bone without horn covering), the horns of rhinoceroses (thickly, matted hair that has keratin but lacks a bony core), and the ossicones of giraffes (skin-covered bony knobs formed from ossified cartilage). Horns provide a diversity of important functions for the animals bearing them, such as defence from predators, tools in fighting other members of the species for territory or mating, feeding, courtship displays, and cooling. Humans also have utilized them for such valued purposes as making musical instruments, carrying items (gunpowder and drink), Traditional Chinese Medicine (TCM), and for making tools, furniture, and decorations. However, sometimes animals have been hunted just for their horns, such as in TCM or selectively hunting bighorn sheep with huge horns as hunting trophies. Such practices have had a detrimental impact on those animals that stand out because of their horns.

True horns are composed of a bony core covered by keratinous material. Bones are semi-rigid, porous, mineralized structures that consist of cells in a hard matrix and that form part of the endoskeleton of vertebrates. Although externally they may appear to be solid structures, in reality they are composed of living bone tissue interlaced with blood vessels, nerve fibres, and so forth. Keratins are tough and insoluble fibrous structural proteins that form the chief, hard, nonmineralized structural component of horns, hair, wool, nails, claws, hooves, and other vertebrate tissues (as well as part of various invertebrate structures). Only chitin rivals the various keratins in terms of their toughness as biological materials.

True horns are found only among the ruminant artiodactyls (even-toed ungulates), in the families Antilocapridae (pronghorn) and Bovidae (cattle, goats, antelope, and so on). These animals have one or occasionally two pairs of horns, which usually have a curved or spiral shape, often with ridges or fluting. In many species only the males have horns. Horns start to grow soon after birth, and continue to grow throughout the life of the animal, except in pronghorns, which shed the outer kertinous layer annually, but retain the bony core. Unlike the horns of the family Bovidae, the horn sheaths of the pronghorn are branched, each sheath possessing a forward pointing tine (hence the name pronghorn).

Similar growths on other parts of the body are not usually called horns, but spurs, claws, or hoofs.

Hornlike Growths

The term "horn" is also popularly applied to other hard and pointed features attached to the head of animals in various other families.

Giraffidae

Giraffes have one or more pairs of bony bumps on their heads, called ossicones. These are covered with furred skin, and although they look as if they ought to have horns on them, they do not.

Cervidae

Most deer have antlers, which are not true horns. When fully developed, antlers are dead bone without a horn (keratin) or skin covering; they are borne only by adults (usually males) and are shed and regrown each year.

Rhinocerotidae

Rhinocerouses have one or two projections on the upper snout that are commonly referred to as horns. These "horns" are not true horns, but are epidermal derivatives, composed of a solid mass of thickly matted hair—keratin—that grows from the skull without skeletal support. That is, these projections are made of keratin and grow continuously, but do not have a bone core.

Ceratopsidae

The "horns" of the ceratopsians like Triceratops were extensions of its skull bones although debate exists over whether they had a keratin covering.

Monodontidae

Male narwhals have a single long tusk, a modified tooth, which looks like a horn, and is twisted like that of the fictional unicorn.

Insects

Some insects (such as rhinoceros beetles) have horn-like structures on the head or thorax (or both). These are pointed outgrowths of the hard chitinous exoskeleton. Some (such as stag beetles) have greatly enlarged jaws, also made of chitin. Members of the family Acrididae are known as the "short-horned grasshoppers" and those of the family Tettigoniidae are known as the "short-horned grasshoppers" because they typically have antennae shorter or longer than their bodies, respectively.

The Horned lizards (Phrynosoma) have projections on their heads consisting of a hard keratin covering over a bony core, similar to mammalian horns.

Many mammal species in various families have tusks, which often serve the same functions as true horns, but are in fact oversize teeth. These include the Moschidae (musk deer, which are ruminants), Suidae (wild boars), Proboscidea (elephants), Monodontidae (narwhals), and Odobenidae (walruses).

Polled animals or pollards are those of normally-horned (mainly domesticated) species whose horns have been removed, or which have not grown. In some cases, such animals have small horny growths in the skin where their horns would be normally–these are known as scurs.

Animal uses of Horns

Animals utilize their horns in a variety of ways. They may be used in fighting, whether for defence from predators or for fighting members of their own species for territory, dominance, or mating priority. Horns may be used in feeding, such as to root in the soil or to strip bark from trees. Some animals with true horns use them for cooling, the blood vessels in the bony core allowing the horns to function as a radiator.

In animal courtship, many animals use horns in displays. For example, the male blue wildebeest (Connochaetes taurinus)andmash;an antelope that is characterized in both sexes by horns that are smooth and initially curving downward or outward—reams the bark and branches of trees to impress the female and lure her into his territory.

Human Uses of Horns

A shofar made from the horn of a Greater kudu, a woodland antelope, in the Yemenite Jewish style. Horns have been used by people for many purposes throughout history. Some bovid horns are used as musical instruments, for example the shofar, which is often made from the horn of a ram (although any animal except a cow or calf can be used) and is used for Jewish religious purposes. The shofur is referenced as far back as the book of Exodus. These have evolved into brass instruments in which, unlike the trumpet, the bore gradually increases in width through most of its length—that is to say, it is conical rather than cylindrical. These are called horns, though today they are made of metal.

Drinking horns' are bovid horns removed from the bone core, cleaned, and polished and used as drinking vessels.

Powder horns were originally bovid horns fitted with lids and carrying straps, used to carry gunpowder. Powder flasks of any material may be referred to as powder horns.

Antelope horns are used in traditional Chinese medicine.

The keratin (horn), sometimes including keratin from other parts of animals, such as hoofs, may be used as a material in tools, furniture, and decoration, among other uses. In these applications, horn is valued for its hardness, and it has given rise to the expression hard as horn. Horn is somewhat thermoplastic and (like tortoiseshell) was formerly used for many purposes where plastic would now be used. Horn also may be used to make glue. Horn bows are bows made from a combination of horn, sinew, and usually wood. These materials allow more energy to be stored in a short bow than wood would.

"Horn" buttons are usually made from deer antlers, not true horn. Use of animal horns is controversial, especially if the animal was specifically hunted for the horn as a hunting trophy or object of decoration or utility. Some animals are threatened or endangered to reduced populations partially from pressures of such hunting.

RESPIRATION IN TERRESTRIAL VERTEBRATES

Lungs are the internal respiratory organs of amphibians, reptiles, birds, and mammals. The lungs, paired invaginations located in one area of the body, provide a large, thin, moist surface for gas exchange. Lungs work with the circulatory system, which transports oxygen from inhaled air to all tissues of the body. The circulatory system also transports carbon dioxide from body cells to the lungs to be exhaled. The process of inhaling and exhaling is calledpulmonary ventilation.

Besides these similarities, there is a great variety in the respiratory systems of terrestrial vertebrates. Frogs, for instance, have balloon-like lungs that do not have a very large surface area. Diffusion across the frog's moist skin supplements the gas exchange through the lungs. Birds have about eight thin-walled air sacs attached to their lungs. The air sacs take up space in the entire bodyThe human respiratory system. Illustration by Hans & Cassidy. Courtesy of Gale Group.

Cavity and in some of the bones. When birds inhale, air passes through a tube called the bronchus and enters the air sacs located in the posterior (rear) of the animal. At the same time, air in the lungs moves forward to air sacs located in the anterior (front). When birds exhale, the air from the anterior air sacs moves to the outside, while air from the posterior sacs moves into the lungs. This efficient system moves air forward through the lungs both when the bird inhales and exhales. Blood in the capillaries of the lungs flows against the air current, which again increases respiratory efficiency. Birds are capable of flying at high altitudes, where the air has a low oxygen content, because of these adaptations of the respiratory system.

CIRCULATORY SYSTEM

The circulatory system consists of blood, a heart, and blood vessels.

Functions of the Circulatory System

The circulatory system functions with other body systems to provide the following:

Transport of Materials

Gasses transported: Oxygen is transported from the lungs to the cells. CO_2 (a waste) is transported from the cells to the lungs.

Transport other nutrients to cells - For example, glucose, a simple sugar used to produce ATP, is transported throughout the body by the circulatory system. Immediately after digestion, glucose is transported to the liver. The liver maintains a constant level of glucose in the blood. Transport other wastes from cells - For example, ammonia is produced as a result of protein digestion. It is transported to the liver where it is converted to less toxic urea. Urea is then transported to the kidneys for excretion in the urine.

Transport hormones - Numerous hormones that help maintain constant internal conditions are transported by the circulatory system.

Contains cells that fight infection

Helps stabilize the pH and ionic concentration of the body fluids.

It helps maintain body temperature by transporting heat. This is particularly important in homeothermic animals such as birds and mammals.

Large Animals

Small animals may not need a circulatory system because the interior cells are close to the surface. Oxygen absorbed from the environment by surface cells can diffuse to interior cells. Wastes produced by interior cells move a short distance to the surface and diffuse into the environment.

Most invertebrates and all vertebrates have interior cells that are too far from the body surface to exchange substances efficiently. They require a circulatory system.

A circulatory system is not needed in small, flat, or porous animals because they have a high *surface-to-volume ratio* and can obtain sufficient absorption directly through their skin.

Gas Exchange and Transport in Invertebrates

Only coelomate animals have a circulatory system.

The choanocytes of sponges use cilia to move water through pores in the sides so that water brings each of the cells all of the nutrients necessary for survival.

Cnidarians have a gastrovascular cavity that provides inner cells with exposure to water. They are only two cell layers thick, so that all cells are exposed to the water for nutrient and gas exchange.

Flatworms also have a gastrovascular cavity to provide for internal cells. Their small size and flattened shape gives them a higher surface-to-volume ratio for better absorption from the environment. Fluid contained within the body cavity of pseudocoelomate animals functions to transport nutrients and wastes but these animals do not have a heart or blood vessels.

Echinoderms have gills on the surface of their skin for gas exchange. Nutrients are distributed by coelomic fluid. Amoeboid cells within the coelomic cavity transport some wastes. The water vascular system functions to operate the tube feet.

Open Circulatory System

In an open circulatory system, blood is pumped from the heart through blood vessels but then it leaves the blood vessels and enters body cavities, where the organs are bathed in blood, or sinuses (spaces) within the organs.

Blood flows slowly in an open circulatory system because there is no blood pressure after the blood leaves the blood vessels. The animal must move its muscles to move the blood within the spaces.

In a closed system, blood remains within blood vessels, pressure is high, and blood is therefore pumped faster.

Arthropods and most mollusks (except cephalopods: nautilus, squid, octopus) have an open circulatory system.

Insects

The coelom of insects has been reduced to a cavity that carries blood (hemolymph). It is called a hemocoel..

dorsal heart —> aorta —> hemocoel

Ostia (openings in the heart) close when heart contracts. When heart relaxes, the ostia open and blood is sucked into openings.

The blood of insects is colorless because it lacks *respiratory pigments*; it functions to carry nutrients, not gases.

Animals with open circulatory systems generally have limited activity due to limitations in the oxygen delivery capability of the system. Insects are able to be active because gas exchange is via a tracheal system.

Closed Circulatory System

In a closed circulatory system, blood is not free in a cavity; it is contained within blood vessels. Valves prevent the backflow of blood within the blood vessels.

This type of circulatory system is found in vertebrates and several invertebrates including annelids, squids and octopuses.

The blood of animals with a closed circulatory system usually contains cells and plasma (liquid). The blood cells of vertebrates contain hemoglobin.

Earthworms

Earthworms have a dorsal and ventral blood vessel that runs the length of the animal. Branches from these vessels are found in each segment.

There are five vessels that pump blood from the dorsal vessel to the ventral vessel.

Earthworms have red blood (due to the pigment hemoglobin) but they have no cells. Hemoglobin binds with oxygen to carry it to the tissues.

RESPIRATORY SYSTEM

The respiratory system (or ventilatory system) is a biological system consisting of specific organs and structures used for the process of respiration in an organism. The respiratory system is involved in the intake and exchange of oxygen and carbon dioxidebetween an organism and the environment.

In air-breathing vertebrates, respiration takes place in the respiratory organs calledlungs. The passage of air into the lungs to supply the body with oxygen is known asinhalation, and the passage of air out of the lungs to expel carbon dioxide is known asexhalation; this process is collectively called breathing or ventilation.

In humans and other mammals, the anatomical features of the respiratory system include includetrachea, bronchi, bronchioles, lungs, and diaphragm. Molecules of oxygen and carbon dioxide are passively exchanged, by diffusion, between the gaseous external environment and the blood. This exchange process occurs in the alveoli air sacs in the lungs.

In fish and many invertebrates, respiration takes place through the gills. Other animals, such as insects, have respiratory systems with very simple anatomical features, and inamphibians even the skin plays a vital role in gas exchange. Plants also have respiratory systems but the directionality of gas exchange can be opposite to that in animals. The respiratory system in plants also includes anatomical features such as holes on the undersides of leaves known as stomata.

COMPARATIVE ANATOMY AND PHYSIOLOGY

Horses

Horses are obligate nasal breathers which means that they are different from many other mammals because they do not have the option of breathing through their mouths and must take in oxygen through their noses.

Elephants

The elephant is the only animal known to have no pleural space. Rather, the parietal and visceral pleura are both composed of denseconnective tissue and joined to each other via loose connective tissue. This lack of a pleural space, along with an unusually thickdiaphragm, are thought to be evolutionary adaptations allowing the elephant to remain underwater for long periods of time while breathing through its trunk which emerges as a snorkel.

Birds

The respiratory system of birds differs significantly from that found in mammals, containing unique anatomical features such as air sacs. The lungs of birds also do not have the capacity to inflate as birds lack a diaphragm and a pleural cavity. Gas exchange in birds occurs between air capillaries and blood capillaries, rather than in alveoli.

Reptiles

The anatomical structure of the lungs is less complex in reptiles than in mammals, with reptiles lacking the very extensive airway tree structure found in mammalian lungs. Gas exchange in reptiles still occurs in alveoli however, reptiles do not possess a diaphragm. Thus, breathing occurs via a change in the volume of the body cavity which is controlled by contraction of intercostal muscles in all reptiles except turtles. In turtles, contraction of specific pairs of flank muscles governs inspiration or expiration.

Amphibians

Both the lungs and the skin serve as respiratory organs in amphibians. The skin of these animals is highly vascularized and moist, with moisture maintained via secretion of mucusfrom specialized cells. While the lungs are of primary importance to breathing control, the skin's unique properties aid rapid gas exchange when amphibians are submerged in oxygen-rich water.

Fish

In most fish, respiration takes place through gills. Lungfish, however, do possess one or two lungs. Thelabyrinth fish have developed a special organ that allows them to take advantage of the oxygen of the air.

3

Class Aves

Birds are vertebrates with feathers, modified for flight and for active metabolism. Birds are a monophyletic lineage, evolved once from a common ancestor, and all birds are related through that common origin. There are a few kinds of birds that don't fly, but their ancestors did, and these birds have secondarily lost the ability to fly. Modern birds have traits related to hot metabolism, and to flight:

- horny beak, no teeth
- large muscular stomach
- feathers
- large yolked, hard-shelled eggs. The parent bird provides extensive care of the young until it is grown, or gets some other bird to look after the young.
- strong skeleton

There are about 30 orders of birds, about 180 families, and about 2,000 genera with 10,000 species. Most of them don't live in Michigan, though there are about 400 species that do.

CHARACTERISTICS OF CLASS AVES

Characteristics of Class Aves are given below:

1) The members of class aves are commonly known as birds.
2) They are worm-blooded with an exoskeleton of feathers.
3) Body usually spindle shaped, with four divisions: head, neck, trunk, and tail; neck disproportionately long for balancing and food gathering.
4) Limbs paired; forelimbs usually modified for flying; posterior pair variously adapted for perching, walking, and swimming; foot with four toes.
5) Epidermal covering of feathers and leg scales.
6) Thin integument of epidermis and dermis.

7) No sweat glands.
8) Oil or preen gland at base of tail.
9) Pinna of ear rudimentary.
10) Fully ossified skeleton with air cavities.
11) Skull bones fused with one occipital condyle.
12) Each jaw covered with a keratinized sheath, forming a beak.
13) No teeth; ribs with strengthening, uncinate processes.
14) Posterior caudal vertebrae reduced and fused as the pygostyle.
15) Pelvic girdle a synsacrum.
16) Aerythrocytes sternum usually well developed with keel.
17) Single bone in middle ear.
18) Fused bones in pelvis, feet, hands, and head
19) Lightweight bones (bones that are either hollow or spongy/strutted)
20) Endothermic
21) Possess a four-chambered heart and in general exhibit high metabolic rates
22) Adept navigational abilities in many species
23) Extraordinary communication and song production
24) Nervous system well developed, with 12 pairs of cranial nerves and brain with large cerebellum and optic lobes. Circulatory system consists of four-chambered heart with two atria and two ventricles.
25) Sexes separate; testes paired, with the vas deferens opening into the cloaca.
26) Females have left ovary and oviduct only.
27) Prental care is well developed.

EVOLUTION AND CLASSIFICATION

The first classification of birds was developed by Francis Willughby and John Ray in their 1676 volume *Ornithologiae*. Carolus Linnaeus modified that work in 1758 to devise thetaxonomic classification system currently in use. Birds are categorised as the biological class Aves in Linnaean taxonomy. Phylogenetic taxonomy places Aves in the dinosaurclade Theropoda.

Definition

Aves and a sister group, the clade Crocodilia, contain the only living representatives of thereptile clade Archosauria. During the late 1990s, Aves was most commonly definedphylogenetically as all descendants of the most recent common ancestor of modern birds and *Archaeopteryx lithographica*. However, an earlier definition proposed by Jacques Gauthier gained wide currency in the 21st century, and is used by many scientists including adherents of the Phylocode system. Gauthier defined Aves to include only the modern bird groups, the crown group. This was done by excluding most groups known only from fossils, and assigning them, instead, to the Avialae, in part to avoid the uncertainties about the placement of *Archaeopteryx* in relation to animals traditionally thought of as theropod dinosaurs.

Gauthier identified four conflicting ways of defining the term "Aves", which is a problem because the same biological name is being used four different ways. Gauthier proposed a solution, number 4 below, which is to reserve the term Aves only for the crown group, the last common ancestor of all living birds and all of its descendants. He assigned other names to the other groups.

The birds' phylogenetic relationships to major living reptile groups.

1. Aves can mean those advanced archosaurs with feathers (alternately Avifilopluma)
2. Aves can mean those that fly (alternately Avialae)
3. Aves can mean all reptiles closer to birds than to crocodiles (alternatelyAvemetatarsalia [=Panaves])
4. Aves can mean the last common ancestor of all the currently living birds and all of its descendants (a "crown group"). (alternately Neornithes)

Under the fourth definition *Archaeopteryx* is an avialan, and not a member of Aves. Gauthier's proposals have been adopted by many researchers in the field of paleontology and bird evolution, though the exact definitions applied have been inconsistent. Avialae, initially proposed to replace the traditional fossil content of Aves, is often used synonymously with the vernacular term "bird" by these researchers.

Most researchers define Avialae as branch-based clade, though definitions vary. Many authors have used a definition similar to "alltheropods closer to birds than to *Deinonychus*." Avialae is also occasionally defined as an apomorphy-based clade (that is, one based on physical characteristics). Jacques Gauthier, who named Avialae in 1986, re-defined it in 2001 as all dinosaurs that possessed feathered wings used in flapping flight, and the birds that descended from them.

Dinosaurs and the origin of birds

Based on fossil and biological evidence, most scientists accept that birds are a specialized subgroup of theropod dinosaurs. More specifically, they are members ofManiraptora, a group of theropods which includes dromaeosaurs and oviraptorids, among others. As scientists have discovered more nonavian theropods closely related to birds, the previously clear distinction between nonbirds and birds has become blurred. Recent discoveries in the Liaoning

Province of northeast China, which demonstrate many small theropod dinosaurs had feathers, contribute to this ambiguity.

The consensus view in contemporary paleontology is that the birds, or avialans, are the closest relatives of the deinonychosaurs, which include dromaeosaurids andtroodontids. Together, these form a group called Paraves.

Some basal members of this group, such as *Microraptor*, have features which may have enabled them to glide or fly. The most basal deinonychosaurs were very small. This evidence raises the possibility that the ancestor of all paravians may have been arboreal, have been able to glide, or both. Unlike *Archaeopteryx* and the non-avian feathered dinosaurs, who primarily ate meat, recent studies suggest that the first birds were herbivores.

The Late Jurassic *Archaeopteryx* is well known as one of the first transitional fossils to be found, and it provided support for the theory of evolution in the late 19th century.*Archaeopteryx* was the first fossil to display both clearly traditional reptilian characteristics: teeth, clawed fingers, and a long, lizard-like tail, as well as wings with flight feathers identical to those of modern birds. It is not considered a direct ancestor of modern birds, though it is possibly closely related to the real ancestor.

Alternative scientific theories and controversies

Early disagreements on the origins of birds included whether birds evolved from dinosaurs or more primitive archosaurs. Within the dinosaur camp, there were disagreements as to whether ornithischian or theropod dinosaurs were the more likely ancestors. Although ornithischian (bird-hipped) dinosaurs share the hip structure of modern birds, birds are thought to have originated from the saurischian (lizard-hipped) dinosaurs, and therefore evolved their hip structure independently. In fact, a bird-like hip structure evolved a third time among a peculiar group of theropods known as the Therizinosauridae.

A small minority of researchers, such as paleornithologist Alan Feduccia of the University of North Carolina, oppose the majority view, contending that birds are not dinosaurs, but evolved from early reptiles like *Longisquama*.

EARLY EVOLUTION OF BIRDS

The earliest known bird (avialan) fossils currently known hail from theTiaojishan Formation of China, which has been dated to the late Jurassicperiod (Oxfordian stage), about 160 million years ago. The avialan species from this time period include *Anchiornis huxleyi*, *Xiaotingia zhengi*, and*Aurornis xui*. The well-known early avialan, *Archaeopteryx*, dates from slightly later Jurassic rocks (about 155 million years old) from Germany. Many of these early avialans shared unusual anatomical features that may be ancestral to modern birds, but were later lost during bird evolution. These features include enlarged claws on the second toe which may have been held clear of the ground in life, and long feathers or "hind wings" covering the hind limbs and feet, which may have been used in aerial maneuvering.

Avialans diversified into a wide variety of forms during the Cretaceous Period. Many groups retained primitive characteristics, such as clawed wings and teeth, though the latter were lost independently in a number of bird groups, including modern birds (Neornithes). While the earliest forms, such as *Archaeopteryx* and *Jeholornis*, retained the long bony tails of their ancestors, the tails of more advanced birds were shortened with the advent of the pygostyle bone in the clade Pygostylia. In the late Cretaceous, around 95 million years ago, the ancestor of all modern birds also evolved better olfactory senses.

Early Diversity

The first large, diverse lineage of short-tailed birds to evolve were the enantiornithes, or "opposite birds", so named because the construction of their shoulder bones was in reverse to that of modern birds. Enantiornithes occupied a wide array of ecological niches, from sand-probing shorebirds and fish-eaters to tree-dwelling forms and seed-eaters. While they were the dominant group of land birds during the Cretaceous period, enantiornithes became extinct along with many other dinosaur groups at the end of theMesozoic era. Many species of the second major bird lineage to diversify, the Euornithes ("true birds", including the ancestors of modern birds), were semi-aquatic and specialized in eating fish and other small aquatic organisms. Unlike the enantiornithes, which dominated land-based and arboreal habitats, most early euornithes lacked perching adaptations and seem to have included shorebirds, waders, and swimming and diving species. The later included the superficially gull-like genus *Ichthyornis* (fish birds), theHesperornithiformes, which became so well adapted to hunting fish in marine environments that they lost the ability to fly and became primarily aquatic. The early euornithes also saw the development of many traits associated with modern birds, like strongly keeled breastbones, toothless, beaked portions of their jaws (though most non-avian euornithes retained teeth in other parts of the jaws). Euornithes also included the first birds to develop true pygostyle and a fully mobile fan of tail feathers, which may have replaced the "hind wing" as the primary mode of aerial maneuverability and braking in flight.

Diversification of modern birds

All modern birds lie within the crown group Neornithes (alternately Aves), which has two subdivisions: the Palaeognathae, which includes the flightless ratites (such as the ostriches) and the weak-flying tinamous, and the extremely diverseNeognathae, containing all other birds. These two subdivisions are often given therank of superorder, although Livezey and Zusi assigned them "cohort" rank. Depending on the taxonomic viewpoint, the number of known living bird species varies anywhere from 9,800 to 10,050.

Due largely to the discovery of *Vegavis*, a late Cretaceous neognath member of theduck lineage, Neornithes is now known to have split into several modern lineages by the end of the Mesozoic era. The earliest divergence from the remaining Neognathes was that of the Galloanserae, the superorder containing the Anseriformes (ducks, geese, swans and screamers) and theGalliformes (the pheasants, grouse, and their allies, together with the mound builders and the guans and their allies). The dates for the splits are much debated by scientists. The Neornithes are agreed to have evolved in the Cretaceous, and the split between the Galloanseri from other Neognathes occurred before the Cretaceous–Paleogene extinction event, but there are different opinions about whether the radiation of the remaining Neognathes occurred before or after the extinction of the other dinosaurs. This disagreement is in part caused by a divergence in the evidence; molecular dating suggests a Cretaceous radiation, while fossil evidence supports aCenozoic radiation. Attempts to reconcile the molecular and fossil evidence have proved controversial.

Classification of modern bird orders

The classification of birds is a contentious issue. Sibley andAhlquist's *Phylogeny and Classification of Birds* (1990) is a landmark work on the classification of birds, although it is frequently debated and constantly revised. Most evidence seems to suggest the assignment of orders is accurate, but scientists disagree about the relationships between the orders themselves; evidence from modern bird anatomy, fossils and DNA have all been brought to bear on the problem, but no strong consensus has emerged. More recently, new fossil and molecular evidence is providing an increasingly clear picture of the evolution of modern bird orders.

This is a list of the taxonomic orders in the subclass Neornithes, or modern birds. This list uses the traditional classification (the so-calledClements order), revised by the Sibley-Monroe classification. The list of birds gives a more detailed summary of the orders, including families.

Subclass Neornithes

Superorder Palaeognathae: The name of the superorder is derived from *paleognath*, the ancient Greek for "old jaws" in reference to the skeletal anatomy of the palate, which is described as more primitive and reptilian than that in other birds. The Palaeognathae consists of two orders that comprise 49 existing species.

- Struthioniformes—ostriches, emus, kiwis, and allies
- Tinamiformes—tinamous

Superorder Neognathae: The superorder Neognathae comprises 27 orders that have a total of nearly 10,000 species. The Neognathae have undergone adaptive radiation to produce the staggering diversity of form (especially of the bill and feet), function, and behaviour that are seen today.

The orders comprising the Neognathae are:

- Anseriformes—waterfowl
- Galliformes—fowl
- Charadriiformes—gulls, button-quails, plovers and allies
- Gaviiformes—loons
- Podicipediformes—grebes
- Procellariiformes—albatrosses, petrels, and allies
- Sphenisciformes—penguins
- Pelecaniformes—pelicans and allies
- Phaethontiformes—tropicbirds
- Ciconiiformes—storks and allies
- Cathartiformes—New World vultures
- Phoenicopteriformes—flamingos
- Falconiformes—falcons, eagles, hawks and allies
- Gruiformes—cranes and allies
- Pteroclidiformes—sandgrouse
- Columbiformes—doves and pigeons
- Psittaciformes—parrots and allies
- Cuculiformes—cuckoos and turacos
- Opisthocomiformes—hoatzin
- Strigiformes—owls
- Caprimulgiformes—nightjars and allies
- Apodiformes—swifts and hummingbirds
- Coraciiformes—kingfishers and allies
- Piciformes—woodpeckers and allies
- Trogoniformes—trogons
- Coliiformes—mousebirds
- Passeriformes—passerines, the songbirds or perching birds

The radically different Sibley-Monroe classification (Sibley-Ahlquist taxonomy), based on molecular data, found widespread adoption in a few aspects, as recent molecular, fossil, and anatomical evidence supported the Galloanserae.

WHAT IS VERTEBRATE RESPIRATORY SYSTEM

Gas exchange across internal gill surfaces is extremely efficient. It occurs as blood and water move in opposite direction on either side of lamellar epithelium. For example the water that passes over a gill first encounters vessels that are transporting blood with low oxygen partial pressure into the body.

Thus oxygen diffuses into the blood water than passes over the vessels carrying blood high in oxygen. More oxygen diffuses inward because this blood still has less oxygen than the surrounding water. Carbon dioxide also diffuses into water because its pressure is higher in the blood than in water. This counter current exchange mechanism provides efficient gas exchange by maintaining a concentration gradient between blood and water over the length of capillary bed.

Respiratory Organs of Frog

Frog can live in water as well as on land. Its larval stages respire by gills, the adult has to develop some special respiratory organs adapted for terrestrial mode of life like other terrestrial vertebrates frog has evolved vascularized paired outgrowths from the lower part of the pharynx known as lungs. They are located inside the body and are simple sac like structures with shallow internal folds that increase the inner surface to form many chambers called alveoli. These are separated from each other through septa. The inner surface of alveoli is attached with blood capillaries. Alveoli are site of exchange of gases. From each lung arises a tube or bronchus. Both bronchi open into larynx or sound box which leads into the buccal cavity through glottis.

Like all other amphibians, in frog, ventilation is a single, two way path. Frog uses positive pressure i.e. it pushes the air into buccal cavity by lowering its bucco pharyngeal floor. During this process it opens the nares and closes the glottis. Then with nostrils closed and glottis opened. Air is pushed into lungs. This is called incomplete ventilation. Air forced into lungs mixes with air already present in lungs and deleted in oxygen. On land this exchange of gases is called pulmonary respiration.

Cirtaneous respiration: When frog goes into water or buries itself in mud, it exchanges gases by its moist and highly vascularized thin skin. This is known as cirtaneous respiration. It can also exchange gasses through its thin vascularized lining of buccal cavity. It is called bucco pharyngeal respiration.

Respiratory system of Bird

Birds are lung breathers. The lungs of a bird are internally subdivided into numerous small, highly vascularized thin membranous channels called parabrochi. In addition to a pair of lungs, a bird has 8 to 9 thin walled non-muscular nonvascular sacs that penetrate the abdomen, neck and even the wings. Air sacs work as bellows that ensure unidirectional flow of air or complete ventilation. Thus a bird must take two breathes to move air completely through the system of air sacs and lungs. First breathe draws fresh air into posterior air sacs of the lungs. The second breathe pushes the first breathe into anterior air sacs and then out of the body. Thus one way flow of air enables a bird to fly at very high attitude without any shortage of oxygen as air coming in lungs is always oxygen rich.

Air exchange in human lungs: Air normally enters and leaves this system through either nasal or oral cavities. From these cavities air moves into the pharynx which is common area for respiratory and digestive tracts. During inhalation air from the larynx moves into the trachea (wind pipe) which branches into right and left bronchus. After each bronchus enters the lungs, it branches into smaller tubes called bronchioles which are part of gas exchange portion of respiratory system. During exhalation intercostals muscles and diaphragm relax allowing the thoracic cavity to return to its original smaller size and increasing the pressure in the thoracic cavity. Abdominal muscles contract pushing the abdominal organs against the diaphragm, further increasing the pressure within the thoracic cavity. The action causes the elastic lungs to contract and compress the air in the alveoli. With this compression alveolar pressure becomes greater than atmospheric pressure, causing air to be expelled from the lungs.

ANATOMY IN INVERTEBRATES

Insects

Most insects breath passively through their spiracles (special openings in the exoskeleton) and the air reaches the body by means of a series of smaller and smaller pipes called 'trachaea' when their diameter is relatively large and 'tracheoles' when their diameter is very small. Diffusion of gases is effective over small distances but not over larger ones, this is one of the reasons insects are all relatively small. Insects which do not have spiracles and trachaea, such as some Collembola, breathe directly through their skins, also by diffusion of gases.

The number of spiracles an insect has is variable between species, however they always come in pairs, one on each side of the body, and usually one per segment. Some of the Diplura have eleven, with four pairs on the thorax, but in most of the ancient forms of insects, such as Dragonflies and Grasshoppers there are two thoracic and eight abdominal spiracles. However in most of the remaining insects there are less. It is at this level of the tracheoles that oxygen is delivered to the cells for respiration. The trachea are water-filled due to the permeable membrane of the surrounding tissues. During exercise, the water level retracts due to the increase in concentration of lactic acid in the muscle cells. This lowers the water potential and the water is drawn back into the cells via osmosis and air is brought closer to the muscle cells. The diffusion pathway is then reduced and gases can be transferred more easily.

Insects were once believed to exchange gases with the environment continuously by the simple diffusion of gases into the tracheal system. More recently, however, large variation in insect ventilatory patterns have been documented and insect respiration appears to be highly variable. Some small insects do demonstrate continuous respiration and may lack muscular control of the spiracles. Others, however, utilize muscular contraction of the abdomen along with coordinated spiracle contraction and relaxation to generate cyclical gas exchange patterns and to reduce water loss into the atmosphere. The most extreme form of these patterns is termeddiscontinuous gas exchange cycles (DGC).

Molluscs

Molluscs generally possess gills that allow exchange of oxygen from an aqueous environment into the circulatory system. These animals also possess a heart that pumps blood which contains hemocyaninine as its oxygen-capturing molecule. Hence, this respiratory system is similar to that of vertebrate fish. The respiratory system of gastropods can include either gills or a lung.

Physiology in mammals

In respiratory physiology, ventilation (or ventilation rate) is the rate at which gas enters or leaves the lung. It is categorized under the following definitions:

Control

Ventilation occurs under the control of the autonomic nervous system from parts of the brain stem, the medulla oblongata and thepons. This area of the brain forms the respiration regulatory center, a series of interconnected brain cells within the lower and middle brain stem which coordinate respiratory movements. The sections are the pneumotaxic center, the apneustic center, and the dorsaland ventral respiratory groups. This section is especially sensitive during infancy, and the neurons can be destroyed if the infant is dropped and/ or shaken violently. The result can be death due to "shaken baby syndrome".

The breathing rate increases with the concentration of carbon dioxide in the blood, which is detected by peripheral chemoreceptors in the aorta and carotid artery and central chemoreceptors in the medulla. Exercise also increases respiratory rate, due to the action ofproprioceptors, the increase in body temperature, the release of epinephrine, and motor impulses originating from the brain. In addition, it can increase due to increased inflation in the lungs, which is detected by stretch receptors.

Inhalation

Inhalation is initiated by the diaphragm and supported by the external intercostal muscles. Normal resting respirations are 10 to 18 breaths per minute, with a time period of 2 seconds. During vigorous inhalation (at rates exceeding 35 breaths per minute), or in approaching respiratory failure, accessory muscles of respiration are recruited for support. These consist of sternocleidomastoid, platysma, and the scalene muscles of the neck. Pectoral muscles and latissimus dorsi are also accessory muscles.

Under normal conditions, the diaphragm is the primary driver of inhalation. When the diaphragm contracts, the ribcage expands and the contents of the abdomen are moved downward. This results in a larger thoracic volume and negative pressure (with respect to atmospheric pressure) inside the thorax. As the pressure in the chest falls, air moves into the conducting zone. Here, the air is filtered, warmed, and humidified as it flows to the lungs.

During forced inhalation, as when taking a deep breath, the external intercostal muscles and accessory muscles aid in further expanding the thoracic cavity. During inhalation the diaphragm contracts.

Exhalation

Exhalation is generally a passive process; however, active or *forced* exhalation is achieved by the abdominal and the internal intercostal muscles. During this process air is forced or *exhaled* out.

The lungs have a natural elasticity: as they recoil from the stretch of inhalation, air flows back out until the pressures in the chest and the atmosphere reach equilibrium.

During forced exhalation, as when blowing out a candle, expiratory muscles including the abdominal muscles and internal intercostal muscles, generate abdominal and thoracic pressure, which forces air out of the lungs.

Gas exchange

The major function of the respiratory system is gas exchange between the external environment and an organism's circulatory system. In humans and other mammals, this exchange facilitates oxygenation of the blood with a concomitant removal of carbon dioxide and other gaseous metabolic wastes from the circulation. As gas exchange occurs, the acid-base balance of the body is maintained as part of homeostasis. If proper ventilation is not maintained, two opposing conditions could occur: respiratory acidosis, a life threatening condition, and respiratory alkalosis.

Upon inhalation, gas exchange occurs at the alveoli, the tiny sacs which are the basic functional component of the lungs. The alveolar walls are extremely thin (approx. 0.2 micrometres). These walls are composed of a single layer of epithelial cells (type I and type II epithelial cells) close to the pulmonary capillaries which are composed of a single layer of endothelial cells. The close proximity of these two cell types allows permeability to gases and, hence, gas exchange. This whole mechanism of gas exchange is carried by the simple phenomenon of pressure difference. When the air pressure is high inside the lungs, the air from lungs flow out. When the air pressure is low inside, then air flows into the lungs.

Immune functions

Airway epithelial cells can secrete a variety of molecules that aid in the defense of lungs. Secretory immunoglobulins (IgA), collectins (including Surfactant A and D), defensins and other peptides and proteases, reactive oxygen species, and reactive nitrogen species are all generated by airway epithelial cells. These secretions can act directly as antimicrobials to help keep the airway free of infection. Airway epithelial cells also secrete a variety of chemokines and cytokines that recruit the traditional immune cells and others to site of infections.

Most of the respiratory system is lined with mucous membranes that contain mucosal-associated lymphoid tissue, which produceswhite blood cells such as lymphocytes.

Metabolic and endocrine functions of the lungs

In addition to their functions in gas exchange, the lungs have a number of metabolic functions. They manufacture surfactant for local use, as noted above. They also contain a fibrinolytic system that lyses clots in the pulmonary vessels. They release a variety of substances that enter the systemic arterial blood and they remove other substances from the systemic venous blood that reach them via the pulmonary artery. Prostaglandins are removed from the circulation, but they are also synthesized in the lungs and released into the blood when lung tissue is stretched. The lungs also activate one hormone; the physiologically inactive decapeptide angiotensin I is converted to the pressor, aldosterone-stimulating octapeptide angiotensin II in the pulmonary circulation. The reaction occurs in other tissues as well, but it is particularly prominent in the lungs. Large amounts of the angiotensin-converting enzyme responsible for this activation are located on the surface of the endothelial cells of the pulmonary capillaries. The converting enzyme also inactivates bradykinin. Circulation time through the pulmonary capillaries is less than one second, yet 70% of the angiotensin I reaching the lungs is converted to angiotensin II in a single trip through the capillaries. Four other peptidases have been identified on the surface of the pulmonary endothelial cells.

Vocalization

The movement of gas through the larynx, pharynx and mouth allows humans to speak, or *phonate*. Vocalization, or singing, in birds occurs via the syrinx, an organ located at the base of the trachea. The vibration of air flowing across the larynx (vocal cords), in humans, and the syrinx, in birds, results in sound. Because of this, gas movement is extremely vital for communication purposes.

Temperature control

Panting in dogs, cats and some other animals provides a means of controlling body temperature. This physiological response is used as a cooling mechanism.

Coughing and sneezing

Irritation of nerves within the nasal passages or airways, can induce coughing and sneezing. These responses cause air to be expelled forcefully from the trachea or nose, respectively. In this manner, irritants caught in the mucus which lines the respiratory tract are expelled or moved to the mouth where they can be swallowed. During coughing, contraction of the smooth muscle narrows the trachea by pulling the ends of the cartilage plates together and by pushing soft tissue out into the lumen. This increases the expired airflow rate to dislodge and remove any irritant particle or mucus.

HUMANS AND MAMMALS

The respiratory system lies dormant in the human fetus during pregnancy. At birth, the respiratory system becomes fully functional upon exposure to air, although some lung development and growth continues throughout childhood. Pre-term birth can lead to infants with under-developed lungs. These lungs show incomplete development of the alveolar type II cells, cells that producesurfactant. The lungs of pre-term infants may not function well because the lack of surfactant leads to increased surface tension within the alveoli. Thus, many alveoli collapse such that no gas exchange can occur within some or most regions of an infant's lungs, a condition termed respiratory distress syndrome. Basic scientific experiments, carried out using cells from chicken lungs, support the potential for using steroids as a means of furthering development of type II alveolar cells. In fact, once a pre-mature birth is threatened, every effort is made to delay the birth, and a series of steroid shots is frequently administered to the mother during this delay in an effort to promote lung growth.

Disease

Disorders of the respiratory system can be classified into four general areas:

- Obstructive conditions (e.g., emphysema, bronchitis, asthma)
- Restrictive conditions (e.g., fibrosis, sarcoidosis, alveolar damage, pleural effusion)
- Vascular diseases (e.g., pulmonary edema, pulmonary embolism, pulmonary hypertension)
- Infectious, environmental and other "diseases" (e.g., pneumonia, tuberculosis, asbestosis, particulate pollutants):

Coughing is of major importance, as it is the body's main method to remove dust, mucus, saliva, and other debris from the lungs. Inability to cough can lead to infection. Deep breathing exercises may help keep finer structures of the lungs clear from particulate matter, etc.

The respiratory tract is constantly exposed to microbes due to the extensive surface area, which is why the respiratory system includes many mechanisms to defend itself and prevent pathogens from entering the body.

Disorders of the respiratory system are usually treated internally by a pulmonologist and Respiratory Therapist.

Plants

Plants use carbon dioxide gas in the process of photosynthesis, and exhale oxygen gas as waste. The chemical equation of photosynthesis is 6 CO_2 (carbon dioxide) and 6 H_2O (water) and that makes 6 O_2 (oxygen) and $C_6H_{12}O_6$ (glucose). What is not expressed in the chemical equation is the capture of energy from sunlight which occurs. Photosynthesis uses electrons on the carbon atoms as the repository for that energy. Respiration is the opposite of photosynthesis. It reclaims the energy to power chemical reactions in cells. In so doing the carbon atoms and their electrons are combined with oxygen forming a gas which is easily removed from both the cells and the organism. Plants use both processes, photosynthesis to capture the energy and respiration to use it.

Plant respiration is limited by the process of diffusion. Plants take in carbon dioxide through holes on the undersides of their leavesknown as stoma or pores. However, most plants require little air. Most plants have relatively few living cells outside of their surface because air (which is required for metabolic content) can penetrate only skin deep. However, most plants are not involved in highly aerobic activities, and thus have no need of these living cells.

ADAPTATIONS FOR EXTERNAL RESPIRATION

1. Primary organs in adult vertebrates are external & internal gills, swim bladders or lungs, skin, & the buccopharyngeal mucosa
2. Less common respiratory devices include filamentous outgrowths of the posterior trunk & thigh (African hairy frog), lining of the cloaca, & lining of esophagus

Respiratory organs:

- Cutaneous respiration
- respiration through the skin can take place in air, water, or both
- most important among amphibians (especially the family Plethodontidae)
- Gills
- Cartilaginous fishes:
    * 5 'naked' gill slits
    * Anterior & posterior walls of the 1st 4 gill chambers have a gill surface (demibranch). Posterior wall of last (5th) chamber has no demibranch.
    * Interbranchial septum lies between 2 demibranchs of a gill arch
    * Gill rakers protrude from gill cartilage & 'guard' entrance into gill chamber
    * 2 demibranchs + septum & associated cartilage, blood vessels, muscles, & nerves = holobranch
- Bony fishes (teleosts):
    * usually have 5 gill slits
    * operculum projects backward over gill chambers
    * interbranchial septa are very short or absent
- Agnathans:
    * 6 - 15 pairs of gill pouches
    * pouches connected to pharynx by afferent branchial (or gill) ducts & to exterior by efferent branchial (or gill) ducts
- Larval gills:
- External gills
    * outgrowths from the external surface of 1 or more gill arches
    * found in lungfish & amphibians
- Filamentous extensions of internal gills

   * project through gill slits
   * occur in early stages of development of elasmobranchs
- Internal gills - hidden behind larval operculum of late anuran tadpoles

Swim bladder & origin of lungs - most vertebrates develop an outpocketing of pharynx or esophagus that becomes one or a pair of sacs (swim bladders or lungs) filled with gases derived directly or indirectly from the atmosphere. Similarities between swim bladders & lungs indicate they are the same organs.

Vertebrates without swim bladders or lungs include cyclostomes, cartilaginous fish, and a few teleosts (e.g., flounders and other bottom-dwellers).

Swim bladders:

- may be paired or unpaired
- have, during development, a pneumatic duct that usually connects to the esophagus. The duct remains open (physostomous) in bowfins and lungfish, but closes off (physoclistous) in most teleosts.
- serve primarily as a hydrostatic organ (regulating a fish's specific gravity)
- gain gas by way of a 'red body' (or red gland); gas is resorbed via the oval body on posterior part of bladder
- may also play important roles in:
- hearing - some freshwater teleosts (e.g., catfish, goldfish, & carp) 'hear' by way of pressure waves transmitted via the swim bladder and small bones called Weberian ossicles
- sound production - muscles attached to the swim bladder contract to move air between 'sub-chambers' of the bladder. The resulting vibration creates sound in fish such as croakers, grunters, & midshipman fish.
- respiration - the swim bladder of lungfish has number subdivisions or septa (to increase surface area) & oxygen and carbon dioxide is exchanged between the bladder & the blood

Lungs & associated structures:

- Larynx
- Tetrapods besides mammals - 2 pair of cartilages: artytenoid & cricoid
- Mammals - paired arytenoids + cricoid + thyroid + several other small cartilages including the epiglottis (closes glottis when swallowing)
- Amphibians, some lizards, & most mammals - also have vocal cords stretched across the laryngeal chamber
- Trachea & syrinx
- Trachea
   * usually about as long as a vertebrates neck (except in a few birds such as cranes)

   * reinforced by cartilaginous rings (or c-rings)
   * splits into 2 primary bronchi &, in birds only, forms the syrinx at that point
- Lungs
- Amphibian lungs
   * 2 simple sacs
   * internal lining may be smooth or have simple sacculations or pockets
   * air exchanged via positive-pressure ventilation
- Reptilian lungs
   * simple sacs in Sphenodon & snakes
   * Lizards, crocodilians, & turtles - lining is septate, with lots of chambers & subchambers
   * air exchanged via positive-pressure ventilation
- Avian lungs - modified from those of reptiles:
   * air sacs (diverticula of lungs) extensively distributed throughout most of the body
   * arrangement of air ducts in lungs ——> no passageway is a dead-end
   * air flow through lungs (parabronchi) is unidirectional
- Mammalian lungs:
   * multichambered & usually divided into lobes
   * air flow is bidirectional:

Trachea <—> primary bronchi <—> secondary bronchi <—> tertiary bronchi <—> bronchioles <—> alveoli

4

The Basics of Vertebrate Evolution

Vertebrates are a well-known group of animals that includes mammals, birds, reptiles, amphibians, and fish. The defining characteristic of vertebrates is their backbone, an anatomical feature that first appeared in the fossil record about 500 million years ago, during the Ordovician period. In this article, we'll look at the various groups of vertebrates in the order in which they evolved to create a picture of how vertebrate evolution unfolded to the present day.

Jawless Fish (Class Agnatha)

The first vertebrates were the jawless fish (Class Agnatha). These fish-like animals had hard bony plates that covered their bodies and as their name implies, they did not have jaws. Additionally, these early fish did not have paired fins. The jawless fish are thought to have relied on filter feeding to capture their food, and most likely would have sucked water and debris from the seafloor into their mouth, releasing water and waste out of their gills.

The jawless fish that lived during the Ordovician period all went extinct by the end of the Devonian period. Yet today there are some species of fish that lack jaws (such as lampreys, and hagfish). These modern day jawless fish are not direct survivors of the Class Agnatha but are instead distant cousins of the cartilaginous fish.

Armored Fish (Class Placodermi)

The armored fish evolved during the Silurian period. Like their predecessors, they too lacked jaw bones but possessed paired fins. The armored fish diversified during the Devonian period but declined and fell into extinction by the end of the Permian period.

Cartilaginous Fish (Class Chondrichthyes)

Cartilaginous fish, better known as sharks, skates, and rays evolved during the Silurian period. Cartilaginous fish have skeletons composed of cartilage, not bone. They also differ from other fish in that they lack swim bladders and lungs.

Bony Fish (Class Osteichthyes)

Members of the Class Osteichthyes first arose during the late Silurian. The majority of modern fish belong to this group (note that some classification schemes recognize the Class Actnopterygii instead of Osteichthyes). Bony fish diverged into two groups, one that evolved into modern fish, the other that evolved into lungfish, lobe-finned fish, and fleshy-finned fish. The fleshy finned fish gave rise to the amphibians.

Amphibians (Class Amphibia)

Amphibians were the first vertebrates to venture out onto land. Early amphibians retained many fish-like characteristics but during the Carboniferous period amphibians diversified. They retained close ties to water though, producing fish-like eggs that lacked a hard protective coating and requiring moist environments to keep their skin damp. Additionally, amphibians underwent larval phases that were entirely aquatic and only the adult animals were able to tackle land habitats.

Reptiles (Class Reptilia)

Reptiles arose during the Carboniferous period and quickly took over as the dominant vertebrate of the land. Reptiles freed themselves from aquatic habitats where amphibians had not. Reptiles developed hard-shelled eggs that could be laid on dry land. They had dry skin made of scales that served as protection and helped retain moisture. Reptiles developed larger and more powerful legs than those of amphibians. The placement of the reptilian legs beneath the body (instead of at the side as in amphibians) enabled them greater mobility.

Birds (Class Aves)

Sometime during the early Jurassic, two groups of reptiles gained the ability to fly and one of these groups later gave rise to the birds. Birds developed a range of adaptations that enabled flight such as feathers, hollow bones, and warm-bloodedness.

Mammals (Class Mammalia)

Mammals, like birds, evolved from a reptilian ancestor. Mammals developed a four-chambered heart, hair covering, and most do not lay eggs and instead give birth to live young (the exception is the monotremes).

FEATHERS AND HAIRS OF VERTEBRATES

Introduction

There is nothing more conspicuous about an organism than its skin. It is our primary means of identifying the organism, and is what defines the boundary of its body. Skin is also the primary means through which an organism interacts with its environment.

Because of its importance as the primary interface between an organism and its environment, the skin is designed to perform many functions.

These functions include:

- Support and protect soft tissues against abrasion, microbes
- Reception and transduction of external stimuli — *i.e.* heat, chemical, tactile
- Transport of materials involved in excretion, secretion, resorption, dehydration, rehydration
- Heat regulation
- Respiration
- Nutrition/nutrient storage — *i.e.* storage of vitamins, synthesis of Vitamin D
- Locomotion
- Coloration — cryptic or display

Different vertebrate taxa have very diverse ways of performing these functions and have evolved many different structures derived from the integument.

Basic Structure of the Integument

The integument consists primarily of the skin and its derivatives. Skin is a functional unit composed layers of fairly distincy epidermis (derived from ectoderm) and dermis (derived from the dermatome of somites) that are separated by the basement membran.

Epidermis

- Is relatively thin in most animals
- The upper layer composed of mostly dead, differentiated cells (stratum corneum) with a lot of keratin which helps the skin maintain some protection against water loss and bacteria
- Continually produced by the most basal layer of the epidermis (stratum germinativum) and consists of cuboidal cells that are generalized and move towards the upper layer as they differentiate
- As the cells move outward, most synthesize keratin, a water-insoluble

protein, the cells become flattened, die, and are sloughed off. Other epidermal cells form multicellular glands or isolated glandular cells.

Dermis

- Is more of a connective tissue than protective
- Irregularly-shaped connective tissue cells that produce the extracellular matrix, including collagen and elastic fibres
- The upper layer (stratum laxum) lies directly below the basement membrane and is mostly loosely-packed cells
- The stratum compactum lies below and contains more tightly-packed cells
- The presence of elastin in the dermis is a synapomorphy of Gnathostomata — in part, the dermis anchors the skin to the underlying musculature
- Also includes dermal scales, blood vessels, nerves, pigment cells, the bases of feathers and hairs, and their associated erector muscles.

Integument of the vertebrate classes

If we again tour through the different taxa that we discussed previously, we find many different forms of integument, based on the different environment that each organism inhabits. Amphioxus has an epidermis with a single layer of cells. A synapomorphy of Craniata is the presence of a stratified (multilayered) epidermis. The horny teeth of lampreys are keratin — most other fishes have little or no keratin in the skin.

There are three major types of hard tissue associated with skin:

Enamel

- The hardest tissue in the body
- Made of hydroxyapatite and has no cells or tubules within it; only about 3% of it is organic
- Ectodermal in origin and is produced by accretion of layers
- Generally it is the most superficial of hard tissues and is found on teeth and the outer layers of denticles, scales and dermal armor - one type of enamel is ganoine

Dentine

- Is softer than enamel and has about 25% organic fibres
- Usually contains tubules occupied by the processes of the mesodermal cells
- Found on the same structures as enamel, but is always deep to the enamel layer

- Some types of dentine are osteodentine, orthodentine, and cosmine, the last of these has characteristic types of canals Bone
- Has about the same level of organic component as dentine
- May have osteons (Haversian systems) as does osteodentine, or may be deposited in layers like orthodentine
- Unlike enamel and dentine, bone may undergo drastic reorganization

Agnathans

The skin of living agnathans lacks dermal bone or scales, but the earliest craniate fossils (Ostracoderms) are known from tiny scales of dermal bone found in the Cambrian period.

These scales had:

- A deepest, thin layer of lamellar bone,
- A thick layer of spongy (vascular) bone,
- Another layer called dentine, and
- A surface coat of enamel-like material, often called ganoine. There was a pore-canal system that likely functioned in electroreception

Chondrichthyes

The skin is covered with denticles or placoid scales with layers of dense lamellar bone, dentine, and enamel Teeth are modified placoid scales

Bony Fishes

Integument of fish is characterized by structures that help the organism maintain its water balance:

- Generally characterized by thin epidermis, with little or no keratinized cells at the stratum corneum
- Mucus secreted from fish's skin which seals out water and also prevents invasion by ectoparasites and fungus
- Glands are unicellular — derived from a single epidermal cell Structures associated most with the fishes are scales:
- Composed of three basic compounds: bone, dentine and enamel (moving from inside to outside); the outside layer, enamel, is the hardest tissue in the body, and therefore can be very protective
- Because they contain compounds that are similar to those in teeth, scales are often compared to teeth
- basal types of scales include:
  - Cycloid scale — thin bony scale having a smooth surface and rounded margins

   - Ctenoid scale — thin bony scale having comblike processes on its outer part and a serrate margin
   - Placoid scale — scaly outgrowth of the skin, that is thicker and more embedded in the skin
   - Cosmoid scale — thick bony plates that are embedded into the skin, that act more like a bony armor
- Perform a more protective function, although the protectiveness of the scale is determined by the thickness of the bone

Amphibians

The earliest tetrapods had dermal scales, which probably functioned as armor. Among living amphibians, caecilians have tiny dermal scales called osteoderms. Their homology with dermal armor is not clear. Amphibians mark the transition between the aquatic and terrestrial environment. Skin remains similar to its aquatic roots and resembles the skin of the fish; however, scales are not present. To prevent water loss, amphibians utilize mucus, which is a similar mechanism that fish use to prevent taking on additional water. However, the mucus in amphibians is secreted by multicellular glands rather than the unicellular glands in fish. Because the integument of amphibians makes them somewhat vulnerable, many amphibians also secrete toxins that prevent them from being eaten by other organisms. The primary gland responsible for the secretion is the parotid gland, located behind the ear of amphibians.

Reptiles

Reptiles show more advanced integumental adaptations to the terrestrial environment because they are more far-removed from the water. In contrast, the cells are more highly keratinized. The integument is modified into horny scales in snakes and lizards. In snakes, the scales on the ventral surface can be further modified into scutes, which can be used in locomotion. In turtles the epidermis is strongly modified into plates that cover the shell, and because they increase in diameter each year, they can be used to age the animals.

Birds

The integument of birds reflects some reptilian ancestry and some new developments of the class. Scales are present on the legs and feet of most birds, and the bill is covered in a tough skin that is highly keratinized. The remaining skin is relatively thin.

The defining characteristic of bird integument is feathers:

- Derived originally from scales, so that scales and feathers are homologous

- Function in flight (flight feathers) as well as temperature regulation (contour feathers)
- Basic structure of feather calamus, rachis and vane, which are derived from a feather follicle. The vane is composed of barbs that help to hold the shape of the feather and can be put back into place during preening.

Birds are not always completely covered in feathers - instead, feathers usually grow along tracts called pterylae, and bare spots are called aptera

Some feathers are modified to perform different functions:

- Down feathers are softer feathers because they lack all the barbs of flight feathers
- Bristles and filoplumes are specially modified feathers that are used in catching prey (*e.g.*, bristles around the bill of swallows and flycatchers) and display (filoplumes of grouse)

Mammals

Mammals generally have skin that conforms to the basic structure described previously, with the epidermal layers of the skin being especially thick in areas such as the soles and the palms of the feet, where proection is needed. Hair is the distinctive characteristic of mammals, and it provides insulation as well as some additional protection to the animal

- Grow in folllicles derived from the stratum germinativum of the epidermus but are rooted in the dermis
- Hair growth continues until the mitosis in the root stops
- Individuals in which mitosis completely stops at the hair root are usually the ones that go bald.

The fine structure of an individual hair consists of three layers: medulla, cortex and cuticular scale (which contain a lot of keratin). (Figure). Softer hairs (such as our fine body hairs) lack a medulla, whereas our scalp hair contains a medulla and is usually very strong. Modifications of hair include guard hairs (that protect the undercoat hair), quills (such as in hedgehogs and porcupines) and vibrissae (the tactile whiskers on the snouts of mammals). Other modifications of mammalian skin includes blubber, which is found in many cetaceans and marine mammals. Blubber is a highly thickened subcutaneous fat layer that adds to the insulation of marine mammals and also acts as a food source for the body.

Glands of the Skin

Glands associated with the skin that help to protect the skin and its associatedd structures, aid in heat regulation, and give off scent.

Include:

- Sebaceous glands which lubricate and waterproof hairs

- Special case in birds the uropygial gland located at the base of the tail which secretes a waxy substance that is used to waterproof and clean feathers.
- two types of sweat glands in mammals aid in heat regulation: eccrine and apocrine sweat glands
- Eccrine sweat glands secrete a watery solution that assists in evaporative cooling on the entire body
- Apocrine sweat glands have thicker secretions that contain more odour, and are sometimes modified into scent glands in some species to use for scent marking (dogs) or defence (skunks); also the wax gland, which secretes the wax in mammalian ears.
- The mammary gland (related to sebaceous glands) which contain fatty tissue in addition to secretory cells that produce milk; usually only become active under hormonal influences, such as the secretion of prolactin by the body that occurs in females during pregnancy and lactation.

Nails, claws, hoofs, horns and antlers: all are integumental derivatives:

- Nails grow from the nail bed located in the epidermis at the distal part of the phalanges; the nail is higly cornified in ungulates whereas in clawed animals the nail is elongated and thickened for defence or predation
- Horns are supported by a bony structure growing out from the skull; surrounding the bony core is a highly keratinized layer of the epidermis which is generally permanent
- Antlers are not present throughout the year, and are shed during the non-breeding season; develop under a protective covering of skin (velvet), which is lost as the antlers mature
- Rhinoceros horns are simply hairlike keratin fibres that are woven together without a bony core - similar to baleen in whales that is used for feeding

Integument Coloration — Pigment Cells

Pigment cells (chromatophores) are derived from neural crest cells that break off from the ectoderm during neural tube formation and are usually found in the dermis.

- In the epidermis of mammals and birds, the pigment cells are usually melanophores which contain the pigment melanin. Melanin is red or blackish brown. Melanophores in the epidermis are usually responsible for slow colour change, such as that related to aging or seasonal changes.

- In groups other than mammals and birds the chromatophores are mostly in the dermis:
- Melanophores are like those of the epidermis iridophores have organelles that contain platelets of guanine pigment, which reflects or scatters light
- Xanthophores and erythrophores have yellowish pteridine pigments and reddish carotenoid pigments
- Dermal chromatophores are responsible for rapid, physiological colour change.

Coloration can be of many types, including cryptic (providing blend into the environment) and aposematic (warning coloriation, that occurs in some snakes)

PROCESSES OF EVOLUTION

One of the most important properties of animals and plants is adaptation to their conditions of existence. Of this obvious but most fundamental point, it will be as well to give a few examples from widely different groups of animals. Birds and mammals both possess a constant temperature which is ordinarily well above that of their surroundings. They should, therefore, be equipped with an arrangement for preventing too rapid loss of heat; and this is provided by the feathers in one case, the hairs in the other.

Swimming birds, almost without exception, are webfooted or lobe-footed. The Jacana, which seeks its food on the floating leaves of water-plants, has enormous toes which distribute its weight as do skis or snow-shoes on the thin crust of snow. Indeed, in general the feet of birds are remarkably well adapted to the life which their owners lead. The egg-eating snake, Dasypeltis, possesses a special mechanism by which it can temporarily dislocate its jaws to swallow an egg whole.

In addition, one of its neck vertebrae bears a downwardly directed spine which protrudes into the gullet, and is used to crack the egg where none of its contents will be wasted. The larvae of crabs live near the surface of the sea, and are provided with long spines which increase friction and make them less ready to sink. But the most striking examples are those known as convergence, where a similar mode of life produces similar effects on quite unrelated animals. For instance, it is advantageous for many animals to be invisible against their surroundings, either to escape their enemies or to approach their prey unobserved. And we find that in polar latitudes many animals are white, at any rate in winter, in deserts many are sandy, in undergrowth many are blotched and streaked so as to break up their outline, and harmonize with the tangle. Others, again, escape their enemies by mimicry, as it is called by resembling other animals which are dangerous or distasteful. Thus wasps, which advertise their sting by their bright black and yellow pattern, are imitated (of course quite unconsciously) by a large number of different sorts of perfectly harmless insects both as regards pattern and bodyform. Ants may be mimicked by many other insects.

Another type of convergence concerns shape. Every one is familiar with the typical fish shape—the stream lines of the body, permitting of rapid motion in water. When other vertebrates have taken to the sea, they too have evolved into a similar shape, as is seen in the Ichthyosaurs, a group of reptiles, and the whales and porpoises, a group of mammals.

This adaptation of every part of an organism to its role, of every whole organism to its mode of life, is universal. It is indeed the direct and most obvious outcome of Natural Selection.

But there is another fact, of perhaps greater importance, which must be taken into account, and that is the existence of higher and lower types of organisms. Within a few square yards we may have the man in a house, the dog in the yard, the worm in a flower-bed, the fish (probably a goldfish) in a pool in the garden, and the amoeba also in the pool, or in a water-butt.

They are all living close together, but their effective environments are amazingly different in extent. Most of the happenings to which the amoeba responds accurately take place within a radius of a millimetre or so; when stimuli like light or vibration affect it, it has no means of discovering anything about the distance or the kind of source from which they come, but responds simply to light or to vibration as such. The total range of environment which the Hydra could conceive of would be a few centimeters each way. The worm is a little larger, but still without any special sense-organs; it can distinguish light from darkness, but cannot see the shape or colour or distance of anything; it can tell when the earth is vibrating, but cannot in any true sense *hear*, because it cannot distinguish tones; it cannot begin to control the environment in the same sort of way as man controls it, because it is not even in contact with most of that environment, locked away from the happenings of the outer world in a windowless existence which is almost incomprehensible to us who are provided with efficient sense-organs. The fish its vision of anything outside the water is very limited owing to the water surface, which if it is rough prevents vision across it. In addition, it is of course confined to water, and so, in this case, to a little pool of a few feet radius.

The dog can see, can hear and smell very well, and can roam over the surface of the earth. Its environment as perceived by its senses is as extensive as that perceived by man's senses; but it cannot understand it in the same way. For instance, we can be pretty sure that no dog could ever come to understand the difference in distance between a cloud, the moon, the sun, and the stars.

In addition, its brain cannot make the same reasoned relations between its sense-impressions as can man.

Although it can and does learn, it can only learn in an unintelligent way, making associations as they come. It does not appear capable of thinking in abstract terms of cause and effect. Its environment must seem both more limited and more chaotic than the man's. The man, if he takes the trouble to be interested in the environment in which he lives, finds it a very marvellous one.

He can get information, by means of a microscope, of invisible things under his nose; he can also obtain information, by spectroscope and telescope and mathematical calculation, about the composition, size, and speed of stars hundreds of light-years away. He can know by letter and newspaper what others are thinking and doing, through history he can enlarge his past far beyond the limits of his single lifetime, and he can make prophecies, sometimes (like those of eclipses and comets) of perfect accuracy, concerning the future.

Also he can relate the different parts of his environment to each other and to general principles. His environment is enormously greater, both in space and time, than the dog's—let alone the amoeba's; and it is much more intelligible.

In addition, the six organisms are of very different sizes and degrees of complication. Amoeba consists of one cell, hydra several thousand, the worm many hundreds of thousands, a man millions of millions. There are many more kinds of cells in man, dog, or fish, than in hydra or even worm. As we ascend the series, we find even greater independence of external forces.

Amoeba and hydra are at the mercy of floods, currents, droughts. A worm is limited to a very small section of the soil. A fish is wide-ranging, and can change its abode at the onset of unfavourable circumstances, but is confined to water. The dog can range on land, and, finally, man is at home in every latitude, and has mastered sea and air as well as earth.

When we look carefully into the matter too the differences between members of the series can be thought of in relation to the environment, but in a much more general way than is the case with special adaptations. Using the word environment to denote the whole series of events and processes with which life can come into contact, and not merely the particular environment of one particular organism, we can say that some animals have more control over the environment than others, and are more independent of it.

The savage is more at the mercy of the elements than is civilized man; civilization is learning to control floods and droughts, to make a pathway of communication of the same sea which to the savage is an impassable barrier. The dog in its turn is not able to cultivate the ground or to kill such varied game as the savage. The fish can exert far fewer different movements than the dog, and is much less able to profit by experience. The earth-worm is not only without specialized organs of locomotion, but also lacks all organs of special sense. So far as independence goes, it is well to remember that amoeba and hydra are bathed, over the whole of their absorptive surface, by the water in which they live, and that accordingly any changes in the composition of that water act immediately upon the vital processes of the animals. In addition, they do not possess any mechanism for regulating their temperature, and so must live slow or fast according as their surroundings are cold or hot.

In man or any mammal or bird, both the temperature and the chemical composition of the fluid which is in contact with all the cells of the body are kept constant with an extraordinary degree of accuracy, and special devices exist for preventing changes in the outer world from exerting their full effect upon the vital processes of the body.

In brief, we may say that high and low organisms can be distinguished by the degree of their control over and their independence of environment. This difference in independence and control is reflected in their structure and their capacity for self-regulation and, in the mental sphere, by the degree of knowledge of the outer world which their sense-organs and brains permit, and in all probability by the intensity of their emotions.

From what we know about evolution it is clear that the highest organisms have developed latest in time. This we can actually see happening as we trace back the history of life in the fossils; it is a probability which amounts for all practical purposes to certainty that the converse is true, and that although the early stages of the earth's history as written in the rocks and fossils are now undecipherable, yet that the first forms in which life appeared were low organisms.

The total thickness since the beginning of the Cambrian is thus about 36 miles. Theincisions indicate the periods when various great mountain-systems were formed—Eurasiatic on the left, American on the right. The dominant forms of animal life areindicated on the right; it will be noted that highly complex forms (Trilobites) had already been evolved in the Cambrian.

At the beginning, then, there were only low organisms, today there exist all gradations between the highest organisms and the lowest undoubtedly many organisms have degenerated during evolution from a higher to a lower condition. If we look back into the history of fossils, and investigate what forms of life were present before and after the development of some new high type, such as man or the mammals, we shall almost always find that the new type has simply been added to the previously existing types.

For instance, the reptiles were the dominant land animals in the secondary period, before the development of the higher or placental mammals; when these were evolved, just before the beginning of the tertiary period, although they speedily became dominant on land, and although many species and even whole sub-groups of reptiles were extinguished, yet the reptilian type as a whole did not perish from off the face of the earth, but continued to exist as well as the mammals.

In the same way, although the advent of man sealed the death-warrant of a great many species of other mammals, reduced the total number of lower mammals very considerably, and deposed them from their previous dominant position, yet lower mammals still exist in abundance, and will undoubtedly continue to do so.

Thus we cannot say that evolution consists simply in the development of higher from lower forms of life; it consists in raising the upper level of organization reached by living matter, while still permitting the lower types of organization to survive. This general direction to be found in evolution, this gradual rise in the upper level of control and independence to be observed in living things with the passage of time, may be called *evolutionary* or *biological progress*. It is obviously of great importance, and can be seen, on reflection, to be another necessary consequence of the struggle for existence.

This improvement has been brought about in two main ways, which we may call *aggregation* and *individuation*. Individuation is the improvement of the separate unit for example, in the series Hydra—Earth-worm—Frog—Man. Aggregation is the joining together of a number of separate units to form a super-unit, as when coral polyps unite to form a colony. This is often followed by division of labour among the various units, which of course is the beginning of individuation for the super-unit, the turning of a mere aggregate into an individual.

Let us take as an obvious example of biological progress the colonization of the land by vertebrates. As a matter of verifiable fact, the sea was already peopled with highly developed fish before the first amphibians appeared on land. In other words, while there existed great competition among vertebrates in the sea, this competition did not as yet exist on the land. Clearly, then, it would be a biological advantage to any species if it were to vary in such a way as to make it able to live on land, for then unchecked multiplication would be possible for it, and it would have fewer enemies. Variations in this direction would thus tend to be preserved; in other words, this particular step in biological progress would be favoured.

As a matter of fact the step is a very large one, and progress was inevitably slow. The Amphibia did not arrive at a complete solution of the problem of leading a terrestrial existence. Their skin is moist, they are usually confined to wet places even in their adult existence, and the earlier part of their life is almost invariably spent actually in water, in the shape of a tadpole larva.

With the evolution of the Amphibia the fringe of the dry land, the territory between land and water, had been conquered, but not the dry land as a whole. Once again, after millions of years during which the Amphibia were the highest vertebrate type, evolving life was confronted with a situation in which a premium was placed upon a further advance: animals born with heritable variations making it possible for them to live farther and more permanently away from water would become heirs of a rich unoccupied territory. Thus it was that the Amphibia, after themselves arising from the fish, in their turn gave origin to the reptiles. In the same way, in the continual struggle that is going on in mammals or birds between herbivore and carnivore, pursuer and pursued, each new advance in speed and size or strength in one party to the conflict, must call forth a corresponding advance in the other, if it is not to go under in the struggle and become extinct. The striking improvement in the running powers of the horse family during its evolution, evinced in increase of size, lengthening of the legs, reduction in the number of the digits, and development of a well-formed hoof, and the similar improvements that occurred in other Ungulates, were accompanied by a corresponding increase in the size, speed, and power of the group of Carnivora during the same geological periods. Each was at the same time the cause and the effect of the other.

A precisely similar state of affairs is often to be seen in the evolution of the tools and weapons and machines of man. For instance, in naval history, the increase throughout the nineteenth century of the range and piercing power of projectiles on the one hand, of the thickness and resistance of armour-plate on the other, provides a very exact parallel with the simultaneous increase of speed and strength in both carnivores and their prey. In Nelson's time, the men-of-war were built of unsheathed wood, and the guns fired round iron balls, with a maximum range of a few hundred yards. To-day, battleships carry 15-inch guns which fling steelcapped and pointed projectiles, laden with high explosive, for a dozen miles or so, while the armour-plating of heavily protected ships may now reach a thickness of a foot or even more of specially-treated nickel steel. The number of digits on both limbs is slowly reduced; the middle digit is enlarged and specialized as a hoof; the fore-arm is strengthened by the fusion of radius and ulna, the ulna in its distal part disappearing; the hind limb is strengthened by the disappearance of the fibula and corresponding enlargement of the tibia; the length of the tooth is increased; its grinding surface becomes more complex.

In addition there is an increase of size and change of proportions. During the interval, progress has been steady and gradual in both departments of naval warfare, each advance in efficiency of guns being the stimulus for new invention in the methods of protection, and vice versa.

It is when there is a general increase of the animal's powers of control that we speak of progress; when the increase is in one special particular only, we speak of specialization. For instance, the horse is specialized for running, the mole for burrowing, the bat for flying, the whale for a marine life, the lion for catching and devouring large animals, the sloth for living in trees.

Each of these animals is well adapted for its particular mode of life, but each is by that very adaptation quite cut off from leading the life of any other. During the late Secondary period there existed similarly specialized types of animals. For example, the Ostrich Dinosaur was adapted for running, the Ichthyosaur for life in the sea, the Tyrannosaur for preying on large animals, the Pterodactyl for flight, and so on.

But all the former list of animals are mammals, all the latter were reptiles. The mammals are higher in that they possess proper temperature-regulation, for instance, and better prenatal care of their young, as well as in many other points. Thus, while the types of specialization, or of adaptation to particular modes of life, are somewhat similar in the two cases, yet each member of the first group is higher than any member of the second, because its general organization is on a more efficient level.

Whenever a new group of animals is evolved, it is found that its members soon become specialized in different directions, thus filling up different vacant places in the economy of Nature. This adaptation to different modes of life, while we call it *specialization* when we are thinking only of one species of animal, is called *adaptive radiation* when we are thinking of the group as a whole. All fish, for instance, breathe by gills, possess fins as limbs, and have other characteristics in common, so that any member of the group can be at once recognized by a brief examination of its structure.

Yet the detailed form of different species of fish is extremely varied. Besides the ordinary type of active, free-swimming fish, like the herring or trout or mackerel, there are fish flattened in adaptation to life on the bottom, some flattened sideways, like the sole and plaice, others flattened from above downwards, like the skates and rays; there are elongated fish like eels and pipe-fish; there are fish with prehensile tails, like the sea-horse; there are the flying-fish adapted for leaping long distances out of the water; mottled fish of irregular outline adapted for living on rocks; deep-sea fish with wonderful phosphorescent organs for searchlights and huge eyes for perceiving the faintest trace of light; cave-fish without eyes at all; sucker-fish adapted for sticking tight to the underside of stones, or for being carried about by larger fish without expending any energy themselves; and so on and so forth through almost every conceivable sort of form possible in an under-water existence. The examples previously mentioned give an idea of some of the adaptive radiation which has taken place in reptiles and in mammals; very similar instances could be taken from any other large group, such, for instance, as the insects. It is interesting to take any such group and to see what part adaptive radiation has played in giving rise to the main sub-groups into which it is divided. There is one particular form of specialization which we have not so far mentioned; and that is the retrograde form of specialization known as degeneration. There are many cases known where animals can be definitely shown to be less highly organized to-day than were their ancestors in the past.

The animal known as Sacculina, for instance, is a parasite upon various sorts of crabs. It consists of a mere bag filled with little else but reproductive cells, and sending out a whole series of branched roots which penetrate the crab's body in all directions and suck nourishment out of it. At first sight, the relationships of this unpleasant creature are very difficult to determine. But when its development is investigated, it is found that it hatches out of the egg as a a free-swimming larva exactly like those found in many Crustacea.

It has jointed limbs, an external skeleton made of chitin, and in fact bears all the distinguishing marks that other Crustacea do, be they crabs or lobsters or shrimps or water-fleas. It is, in fact, a crustacean which has become adapted to a parasitic life; and in so doing it has acquired special adaptations, such as the root-like organs, specially suited to that life, while it has lost sense-organs, limbs, digestive system, and everything else necessary for leading a free and independent existence.

It has lost more than it has gained, its organization has become simpler, its independence less; in fact it has gone down the evolutionary hill, and the direction of its history has been in most ways the opposite of the direction which characterizes biological progress.

The form in which Sacculina hatches out resembles in general many crustacean larvae; but it is particularly like the early stages of the animals known as barnacles. Every one who has been to the seaside knows what an acorn barnacle looks like—a little creature attached to rocks or piles, enclosed in a white shell, and capable of sending out of a slit in the top of the shell a regular sweep-net composed of a number of 'arms'—really appendages—with which it drags minute floating particles of food in towards its mouth.

Even in the adult state a barnacle shows some resemblance to ordinary Crustacea, especially in its jointed limbs and chitinous external skeleton; the early free-swimming larva clinches the matter and gives complete proof, as in the case of Sacculina, of their crustacean nature and affinities. The very close resemblance of the larva of barnacles to that of Sacculina points to an especially close connexion between the two sorts of animals; and as a matter of fact, the two are undoubtedly descended from a common stock.

The barnacle is degenerate as well as the Sacculina, but it is not so degenerate. It still, for instance, possesses organs for capturing and digesting food. On the other hand, it has lost its organs of special sense and of locomotion. Further, it is not adapted to the same general mode of life as is Sacculina; it is not parasitic, but sedentary or sessile.

This settling down and becoming fixed is the other great cause, besides parasitism, of degeneration in animals; as would be expected, the degeneration due to a sedentary life is rarely so great as that due to parasitism, since the sedentary animal does not obtain its food ready-digested as do most parasites.

It must not be supposed that because the general rule among animals is that time brings change, that therefore time *invariably* brings change. The common lamp-shell Lingula, for instance, has persisted without any appreciable change whatever in the structure of its shell for the prodigious period of time, certainly over five hundred million years, which has elapsed since the Cambrian epoch in the Primary period. Even when individual species have changed, the general characters of groups have often persisted with very little modification, as, for instance, those of dragonflies since the coal-measures, of shark-like fish since the Silurian.

In some cases this may mean that for some unknown reason the species or the group has lost the power of varying to any considerable extent. More often, probably, it so admirably fills one particular niche in the order of things, and a niche which stays the same throughout the geological periods, that it pays for the animal or the group of animals to stay as they are, leaving it to other groups or to other branches of the same group to colonize new niches and to progress towards fuller existence.

To sum up, we may say that two main types of evolutionary change result in animals (and also, as a matter of fact, in plants) from the struggle for existence and the constant appearance of inheritable variations. In the first place, once a new type or plan of structure has been evolved, it undergoes adaptive radiation; in other words, there are developed a number of separate species all built on the same general ground-plan, but adapted to different and usually incompatible modes of life. In the second place, new types and new plans are continually appearing as time goes on, and progress is marked by the fact that among the later-evolved types there is to be found greater complexity of organization, greater control and independence of environment, than among the earlier.

It might perhaps be thought that specialization was often the same thing as progress. Specialization, however, implies close adaptation to one particular mode of life, while progress means greater general efficiency. If we look into the actual history of animals in the past, we find that specialization in one direction involves the sacrifice of possible advance in other directions, and is a barrier in the long run to any but a quite limited degree of progress.

As a result of its long course of specialization, extending for tens of millions of years through the better part of the Tertiary epoch, a specialization all tending towards greater efficiency in running and browsing, the horse stock has, it seems, cut itself off from the possibility of adapting itself to other modes of life—to a life in the water or in the trees, or to a carnivorous habit.

There is a limit to the perfection which any line of specialization can attain. While the horse was growing larger, developing hoofs, reducing the number of its digits, another and wholly different type was being evolved in the person of man. If it were not that horses are useful to man, and are accordingly domesticated, they would now be wholly or almost wholly extinct.

The 'natural' enemies of the horse are the large carnivores. These are built on the same general plan as the horse—the mammalian—and indeed are the results of the specialization of the same plan in another direction. The same limits are thus set to them as to the horse stock. Before the advent of man, a state of equilibrium existed between the horse and its enemies, the latter not able to destroy the former entirely, the first not able to escape the payment of some toll to the second.

But the horse came up against the wholly new biological conditions introduced by the new type, man; it was in direct conflict with man's cunning and tools and his habit of hunting in bands; still more important, it had to compete with the indirect effects of the new development, such as the settlement and cultivation and enclosure of the land. Against all this the horse was powerless. It could not develop far enough or fast enough to adapt itself to such sudden changes, and as a result it is becoming extinct as a wild species.

Over and over again the same thing happens, and the specialized representative of the old type, plant or animal, is extinguished in competition with the specialized representative of the new. For tree-like representatives of the horsetail type, which existed in the Carboniferous period, we have seed-bearing trees to-day; for pterodactyls, birds and bats; for dinosaurs, the large mammals; for the large early amphibians, the Stegocephalia, we have crocodiles; for the abundance, both in numbers and in species, of the larger mammals in the Pliocene, we have the multifarious activities of the swarms of man.

The new type seems always to have arisen from some comparatively unspecialized branch of the old, and to have attained its pre-eminence by means of adaptations towards general instead of towards special efficiency. Man has ousted the other mammals from their dominant position owing to the development of his mind. Through his particular type of mind he is able to deal rationally with the problems that confront him; tools, machines, tradition, civilization, and unexampled control and independence have been the result. The human mind is not merely adapted for solving one or two particular problems; it represents a *method* more efficient than any previously adopted for dealing with any and every problem that may confront an organism.

Man's body is not highly specialized, and he seems to have arisen, through a monkey-like ancestor, from some unspecialized early mammalian stock like the Insectivores. Nor did the early mammals show any signs of specialization. All the fossil mammals that we know of during the time when the great period of reptilian dominance and specialization lasted were small, primitive creatures, at first sight not likely to wrest the palm from their powerful rivals.

Very similar chains of events may be seen taking place in the evolution of human machines and inventions. Take, for example, the history of transport. The most primitive method was the carrying of single loads by human beings or pack-animals. After that came the invention of vehicles in general and of wheeled vehicles in particular. The wheeled vehicle became specialized ('adaptive radiation') in innumerable ways.

We have the war-chariot; the rapid vehicle for passenger transport and for pleasure; the heavy wagon and cart; the van and pantechnicon; the four-inhand mail-coaches bowling daily at fixed hours along the main roads of the land. A limit was set to the capacity and the speed of such vehicles by the speed and strength of beasts of burden on the one hand, and by the imperfections of road-surface on the other.

At the beginning of last century a new type of vehicle was evolved for which these limitations no longer existed. It was discovered how to replace the energy of animals by that of steam, and in large part to overcome the difficulties of surface friction by making the wheels of the vehicles run on metal rails. As a result, steam locomotives became for certain purposes the 'dominant type' of vehicle within an extremely short period of time.

Here again the type itself has been improved, so that we have now for some time been close to the limit of its possibilities. It does not seem possible to run a profitable service at speeds of much over sixty miles an hour. About half a century later a new plan was evolved. The internal combustion engine was produced, and gave certain great advantages, notably in being ready to start at once without the long preparation of 'getting steam up'. It appears to be the fate of new types to lead a precarious existence for a considerable time before they can compete successfully with the dominant types of the period.

This was so, as we saw, with the earliest mammals; it was so for the steam-engine; and it was so, to a very marked degree, for the internal combustion engines. They were laughed at when their inventors took them out on the roads; the law laid down that they should be preceded by a man with a red flag; the early defects in construction did actually occasion many a breakdown. But within thirty years they came into their own.

Meanwhile, still another competitor is in the field—the flying machine, a totally new type, abandoning not only the particular device of the wheel, but the whole element to which the wheel was adapted. It looks as if in certain respects, where speed is the main object, the aeroplane would become the dominant type of vehicle; but that it would leave the major part of transport to be dealt with by train and by motor.

Another interesting parallel with the evolution of organisms is found in the fact that although there has been progress, although the dominant type of vehicle has altered with the passage of time, yet many representatives of the old types have survived. They have managed to survive by becoming restricted to a few special conditions and places. Packanimals, for instance, while once universal, are now only employed in mountainous or roadless countries. The American buggy is still of the greatest use over the unmade roads which are still to be found in so many parts of the United States.

The horse-drawn vehicle will long hold its own in businesses in which not much capital is available, and speed and great power are not of the first importance.

This survival of all or almost all the types that arise in evolution, even though new types arise and supremacy changes hands, is of general occurrence in the development of animals and plants, and is at first sight very puzzling. However, the insight which is gained by looking, as we have done, into the evolution of something familiar and human like the means of locomotion, helps us to understand the more complex and slower-moving processes of organic evolution.

Another point which is brought out by the study of the development of some human contrivance such as the means of transport, is the great speed of change now possible in human affairs as against the slowness of change prevailing in lower organisms.

The whole period from the stage-coach to the aeroplane is comprised in well under two centuries. The resulting change in human habits has been enormous; to produce comparable changes in the habits of an animal stock there would be needed a period certainly to be reckoned in tens of thousands, possibly in millions of years.

On the other hand, the evolution of machines is a perfectly real evolution. Two different types of machines capable of performing the same general function—such, for instance, as the motor-lorry and the goods steam-engine—do come into a very real competition with each other, and the issue of the struggle is decided by a form of true natural selection, depending in the long run upon which of the two pays the better.

Here again the study of machines throws light upon the course of events in animals. It is often supposed that evolution must involve some conscious effort on the part of the evolving organism, that the struggle for existence is a conscious struggle, or that a species in some mysterious way 'learns' how to develop some new improvement in its structure.

As a matter of fact, almost the whole of such ideas are purely metaphorical, and arise simply because we read the processes of our own minds into the operations of nature; it is not scientifically correct to speak, for instance, of *purpose* except in relation to human minds. We see at once that the machines have no idea themselves of the direction of their evolution, that the 'struggle' between them is only a metaphorical struggle, that the selection between them is so far as they are concerned a mere sifting process, the issue of which depends upon the advantages or disadvantages which they may happen to possess.

It is the same with organisms. If two races of animals come into competition, the issue is decided by the qualities which each happens to possess; 'natural selection' is a name for the effect exerted by all the forces of the environment with which they come into relation, an effect which acts again like an automatic sieve, and lets some through to perpetuate themselves, keeps back and so extinguishes others.

The 'struggle' and the 'competition' are again usually metaphorical only. For instance, when the common housesparrow was introduced into America, it entered into competition with many of the small sparrow-like birds which had developed in that country. But the struggle did not take the form of a war between the invader and the original inhabitants.

It was an indirect struggle, due to the fact that both lived upon the same sort of food, both occupied the same sort of sites. The European sparrow happened to be endowed with qualities which gave it an advantage, and as a result it has spread enormously over the American continent, while many of the native birds have correspondingly decreased. If we wish to use a human metaphor, we can say that its success has been the result of 'peaceful penetration', not of fighting.

The difference between machines and organisms, of course, is that the machines are directly designed by man, whereas the place of the designer in animal evolution is (roughly) taken by the variation which seems to be universal in organisms. It is variation which provides the raw differences upon which the sieve of natural selection can work.

Mention must also be made of the theories of evolution which are summed up under the term *orthogenesis*, which means evolution in straight lines. It is frequently found, as for instance in the development of the horses or of the elephants, that evolution as revealed by fossils proceeds straight onward through geological time in a perfectly definite direction-in the horses towards single toes and hoofs, in the elephants towards great bulk, tusks, and trunk. Orthogenesis is sometimes used merely as a descriptive term, to denote this observed fact of straight-line evolution.

But by others it is used to mean that there exists some inner necessity for the evolutionary line in question to develop in just that one way and no other. However, a series of fossils, even if beautifully complete, can really give us no insight into the method by which its evolution occurred. Whenever the direction in which the series is evolving is adaptive or seems biologically advantageous, the orthogenetic series can be perfectly well explained by natural selection. On the other hand, there do exist cases where at least no advantage can be perceived by us in the direction pursued.

This is so, for instance, as regards the Ammonites, the extinct cephalopod molluscs which died out near the end of the Secondary period. Near the close of their time on earth, they often evolved orthogenetically into the most bizarre forms, their spiral shell becoming unwound, or irregular. Possibly in such cases a real causal orthogenesis, an inwardly determined mode of variation, is at work. But we are justified in saying that such cases, if they occur at all, are certainly rare.

EVOLUTION OF HEART IN VERTEBRATES

Introduction

Hearts take in oxygen poor blood, pump it through the pulmonary circuit (lungs/gills) where it gets oxygenated, and then they pump it out to the rest of the body. There are many many small veins (take blood to the heart) and arteries (carry it away) which connect the heart to the pulmonary circuit and keep it all running right.

Blood pressure has to be carefully balanced in all tubes so that flow pressure is maintained and blood keeps moving, but not so fast as to explode certain areas or capillaries. One might say this careful and complex hydrostatic wiring might be easily selectable in an evolutionary scenario, but there's no getting around the fact that there's a very complicated and highly balanced network of biological fluid mechanics going on inside a heart than most artificial heart engineers probably care to think about.

Given all this, there are 3 basic ways to make a heart found in animals: a 2 chambered heart, a 3 chambered heart, and a 4 chambered heart. Fish have 2 chambers, one atrium and one ventricle. Amphibians and reptiles have 3 chambers: 2 atria and a ventricle. Crocodiles are the one reptilian exception, as they have 4 chambers (2 atria, 2 ventricles). Birds and mammals have 4 chambers (2 atria and 2 ventricles).

Differences between the Hearts

The fish heart (figure a) is much different than the amphibian/reptile/bird/mammal heart (figures b and c). Hearts are very complex—they're not just a bunch of random arteries and veins connecting tissue. Fish hearts simply draw in deoxygenated blood in a single atrium, and pump it out through a ventricle.

This system is termed "single circulation", as blood enters the heart, gets pumped through the gills and out to the body, Blood pressure is low for oxygenated blood leaving the gills. 3 and 4 chambered hearts have a pulmonary circuit (pathways taking blood from heart to lung and back to heart) that is very complex and must be set up such that blood can travel from the heart to become oxygenated in the lungs and then be properly pumped back the heart and out to the body.

The 3 (and 4) chambered heart has "double circulation" and is quite different from "single circulation" of fishes.

"Double circulation" has an interior circuit within the heart—blood enters the heart, leaves the heart and gets oxygenated, enters the heart again, and then gets pumped out to the body. Because "Double circulation" allows oxygenated blood to be pumped back into the heart before going out to the body, it pumps blood with much more pressure and much more vigorously than "single circulation".

There are 4 steps involved with blood entering the heart:

1. Oxygen poor blood enters the first atrium.
2. Oxygen poor blood is fed to the first ventricle, which pumps it out to the pulmonary circuit (lungs) where it is enriched in oxygen.
3. Oxygen rich blood just leaving the lungs is pumped back into the second atria.
4. Oxygen rich blood is then fed to the second ventricle, which pumps the oxygen rich blood out of the heart and back into the body for usage.

The 4 chambered heart differs from the 3 chambered heart in that it keeps oxygenated blood completely separate from de-oxygnated blood, because there is one ventricle for deoxgynated blood and one for oxygenated blood. In the 3 chambered heart, a single ventricle pumps both out of the heart, and there is some mixing between fresh and old blood.

The 2 ventricle-4 chamber heart prevents mixing allows the blood leaving the heart to have far more oxygen than it would otherwise.

This is good for enhancing the more fast paced lifestyle that birds and mammals tend to have, giving an advantage to having a 4 chambered heart.

Problems with Evolving the Hearts

Getting a Heart, Period

Evolving a 2 chambered heart in the first place is very difficult because the circulatory system is irreducibly complex.

At least 3 subsystems are necessary:

1. An organ for enriching the hemophlegm (blood) with oxygen (lungs/ gills),
2. A complex network of closed tubes to carry the energy-rich blood to the body (veins and arteries), and
3. A pumping mechanism (heart) to transport energy-rich fluid throughout the body. One common way an evolutionist might try to get around problems associated with irreducible complexity might be to imagine scenarios where some or all of the subsystems could originate in a freestanding manner, functional on their own. However in this case what good is a closed tube network without a pumping mechanism to transport fluid, and what good is a pump without the fluid or the tubing? Oxygen exchange occurs in many organisms through the skin without a chordate-like circulatory system, but what is the advantage for such an organism to randomly mutate itself a single oxygen exchange organ (the lungs/gills)? Even so, once a single oxygen exchange organ was in place, it would need the very transport network provided by the heart, veins and arteries. Scenarios attempting the circulatory system in a step by step manner would fail and not take into account the overall complexity of the system.

Even if the subsystems of the circulatory system could be evolved on their own, evolving a freestanding 2 chambered heart on its own would be very difficult, even if it didn't need other components of the circulatory system to be useful.

A human heart is an 11 ounce pumping machine the size of a human fist which beats over 2 billion times and pumps over 100 million gallons of blood over the course of the average human lifetime.

It is primarily a shell with a carefully balanced interior network of holes and valves which keep fluid constantly flowing in, out, and in the right direction everywhere in between.

A large number of fluid-directing parts as well as a very strong and properly and complexly shaped pumping muscle are necessary for it to function. The positioning of the parts of a heart are a good example of specified complexity. For this reason, the heart, as well as the entire circulatory system, are unevolvable in a step by step manner.

Getting a 3 Chambered Heart from a 2 Chambered Heart

Given a 2 chambered heart, experts do not know when, how, or in what lineage the alleged transition from the 2 chamber fish heart to the 3 chambered amphibian heart took place, mainly because this is a very difficult transition to even imagine. A 3 chambered heart has "double circulation" and is irreducibly complex with respect to "double circulation". 2 chambered hearts of fish have "single circulation" and the basic design is very different from a 3 chambered heart.

A vastly oversimplified explanation of the evolution of the heart might be to say that by simply duplicating the some or all of the chambers of a 2 chambered heart, one could easily evolve a functional 3 or 4 chambered heart. Of course the way the evolutionary story goes, fish turned into amphibians which turned into reptiles which became mammals and birds, so the heart is said to have evolved from 2 → 3 → 4 chambers.

A direct 2 → 4 chamber transition is never thought to have taken place and would have even more difficulties than a 2 → 3 chamber transition, so we will focus on the 2 → 3 chamber transition. However, in reality the changes that would need to take place for this transition are far more than merely the duplication of one or more chambers. It could not happen in a step-by-step manner where intermediates are functional.

The 2 chamber heart → 3 chamber heart transition requires much more than the duplication of an atrium, because the interior circuit causing the "double circulation" of the 3 chambered heart must also be created. Duplicating the atrium without a closed circulatory network for "double circulation" would cause the heart to suck nothing but interstitial fluid out of the body.

"Double circulation" only works when there is a loop feeding from a ventricle back to the heart, and back to a ventricle. Thus, both the loop and the new atrium are necessary for a 3 chambered heart to function. And though the loop (interior circuit) sounds simple, it really must be a complex tube network with valves in the right places to keep fluid flowing properly.

Single circulation hearts pump blood directly through the gas exchange organ and out to the body (figure). Double circulation pumps blood to the lung/gills through an "interior circuit" loop and then back to the heart before going out to the body (figure).

And if this new loop doesn't connect with the gas exchange organ, then the new loop is functionless and useless. The easiest way to make this transition happen is probably to have the vein leaving the gas exchange organ feedback to become an artery feeding back into the heart. Of course this means the new vein-artery simultaneously needs the proper valves so it can function like an actual part of the heart itself. Probably of most importance is the fact that there is now no vein leaving the heart and pumping blood back out to the body. In other words, to create this interior circuit of double circulation, a new vein must be created and blood flow out of the heart completely rewired to the rest of the body.

Even if gas exchange organ could be bypassed (making it a less complex, though functionless and not selectively advantageous circuit), the main problem with going from single circulation to double circulation is that somehow this new circuit has to wire itself to the lungs. Either way, the vein leaving the heart must somehow also become an artery fed back into the heart through a new functional atrium and then a new vein must be created so that blood leaving the heart still gets out to the rest of the body through the circulatory network.

Finally, the heart muscle has to adapt to all of these changes, especially such that beating can occur to pump through the new atrium and associated fluid pressure changes.

In other words 4 primary changes are needed to go from single to double circulation:

1. The duplication of the atrium such that fluid transport through new atrium is functional
2. A conversion of the vein leaving the heart into an artery at the other end such that it is fed back into the heart.
3. A complete rewiring of how blood finally leaves the heart and goes out to the rest of the body (the creation of a new vein and rewiring).
4. Modification of the heart muscle to beat properly and accommodate the additional atrium and fluid pressure changes associated with the rewiring.

If any of these steps are missing, double circulation won't work. And this says nothing about the many valves and other smaller veins and arteries associated with double circulation which characterize true hearts as well changes needed in the pumping mechanism of the heart muscle to accommodate a completely new atrium and fluid-pressure balance.

The transition from 2 to 3 chambers requires a change from single to double circulation which involves at least 4 major simultaneous changes including the complete rewiring of how blood leaves the heart to the rest of the body.

Many more minor simultaneous changes associated with mechanics of proper fluid transport would also be necessary. It not possible for double circulation system to evolve from a single circulation heart system in a Darwinian step-by-step manner because too many changes are necessary, making the 3 chambered heart unevolvable from a 2 chambered heart.

Getting a 4 Chambered Heart

With respect to hearts with "double circulation", the 4 chamber mammalian heart probably isn't irreducibly complex. Going from 3 to 4 chambers really doesn't look all that complicated (of course they're still very different and this is totally oversimplified, but I'm only talking about basic organ design). Basically, the single ventricle in the 3-chambered heart is split into 2 chambers in the 4 chambered heart, making 2 ventricles instead of one. A "double circulation" heart could work with 3 chambers—and it does in reptiles and amphibians. The human heart has 2 atrium-ventricle pairs, which beat in succession something like pistons in a car. Only one ventricle is really needed to pump the blood. But just like an 8 cylinder engine put out a lot more horsepower than a 4 cylinder, so does a 4 chamber (2 pairs) heart have a much more power to supply the body with energy and oxygen-rich blood than a 3 chamber (1 ventricle) heart. There is a huge advantage to having a 4 chamber heart. Taking away the additional capillary complexities associated with suddenly having a second ventricle, one might be able to argue that the 3 to 4 chambered heart transition isn't difficult to imagine, relatively speaking, as all one has to do is note the strong advantage of separating oxygenated and de-oxygenated blood and then divide the ventricle in half. Of course the key word here is "imagine".

Irreducible complexity is a real phenomena, and it can be analysed, and so in some cases it might not exist, and in the case of the 4-chambered heart case it probably doesn't. Though the 4 chambered heart may not be irreducibly complex with respect to "double circulation", it might still be the result of intelligent design and not evolution, and irreducible complexity doesn't have to exist in all instances for it to exist in some. Overall, regardless of chambers, the heart has had much design put into it. And the complex double-circulatory heart bears strong marks of intelligent design.

Getting a Bird Hear

Finally, the evolutionary origin of the bird heart-lung system is puzzling for the dino-bird hypothesis because "no lung in any other vertebrate species is known which in any way approaches the avian system.".

Molecular biologist Michael Denton had the following to say about the evolutionary origin of the bird-heart-lung system:

- Just how such a different respiratory system could have evolved gradually from the standard vertebrate design without some sort of direction is, again, very difficult to envisage, especially bearing in mind that the maintenance of respiratory function is absolutely vital to the life of the organism. Moreover, the unique function and form of the avian lung necessitates a number of additional unique adaptations

> during avian development. As H. R. Dunker, one of the world's authorities in this field, explains, because first, the avian lung is fixed rigidly to the body wall and cannot therefore expand in volume and, second, because the small diameter of the lung capillaries and the resulting high surface tension of any liquid within them, the avian lung cannot be inflated out of a collapsed state as happens in all other vertebrates after birth. In birds, aeration of the lung must occur gradually and starts three to four days before hatching with a filling of the main bronchi, air sacs, and parabronchi with air. Only after the main air ducts are already filled with air does the final development of the lung, and particularly the growth of the air capillary network, take place. The air capillaries are never collapsed as are the alveoli of other vertebrate species; rather, as they grow into the lung tissue, the parabronchi are from the beginning open tubes filled with either air or fluid

Denton goes on to say that, "The avian lung brings us very close to answering Darwin's challenge: 'If it could be demonstrated that any complex organ existed, which could not possibly have been formed by numerous, successive, slight modifications, my theory would absolutely break down.' One respiratory physiology expert, John Ruben, critiquing this evolutionary scenario noted that, "a transition from a crocodilian to a bird lung would be impossible, because the transitional animal would have a life-threatening hernia or hole in its diaphragm... It seems clear that a bird's radically different system of breathing, in which air is continuously drawn through its lungs, could not have evolved from the hepatic-piston system we see in this theropod dinosaur." Plus, we're also dealing with convergent evolution here. The 4 chamber system had to have evolved twice (and also in crocodiles), if it evolved at all, because mammals and birds are allegedly from very different reptilian stock. It isn't just a neat case of common ancestry being clear cut by characteristics. There are many other animals with hearts—some worms have like 5 or 6 hearts. These heats all probably evolved independently however. So, there is not a nice neat Darwinian tree with respect to hearts. Common design better explains this, because odds are low you would come up with a similar structure through chance evolution alone.

EVOLUTION OF AORTIC ARCHES AND PORTAL SYSTEMS IN VERTEBRATES

Circulatory System

Vertebrates have the most highly evolved circulatory system in the animal kingdom.

The circulatory system performs a variety of functions including:

- Transport of respiratory gases, nutrients, metabolic wastes, hormones and antibodies
- Maintain internal environment (homeostasis) in conjunction with the kidneys
- Responds quickly to the changes in the body depending on the needs of the moment

The circulatory system is made up of two primary components:

1. *Blood-vascular system*: A closed system composed of the heart, arteries (which distribute blood from the heart to the tissues), veins (return blood from the tissues to the heart) and capillaries (small thin-walled vessels at which physiological exchange occurs) and the blood — although the system itself consists of a continuum of ducts, all are interconnected and allows for little or no loss of contents
2. *Lymphatic system*: Drains fluids that accumulate in the tissues (tissue fluids), which are first collected by lymphatic capillaries, which pass into lymphatic vessels and then empty into the venous system

The circulatory system has more individual variation than any other system and is the first of all organ systems to become functional during development. The system is also highly adaptable — you can graft veins to other locations (such as in a heart bypass) or tie off a vessel without seriously inconveniencing the system

Blood and Blood Vessels

Blood is a fluid tissue containing cellular elements that are derived from mesoderm.

There are two primary components to blood:

1. *Plasma*: Constitutes approximately 2/3 of the blood and is composed of about 90% water
   - The non-water component contains fibrinogen, which contributes to blood clotting, and globulins, which respond to the entry of foreign materials into the body.
2. *Cellular*: Consists of two types of cells
   i. Erythrocytes — the red blood cells which carry hemoglobin that binds oxygen for transport to the tissues
   ii. Leukocytes — the white blood cells that destroy foreign bodies through phagocytosis and are also involved in the immune response

Blood Cells are Produced by Hemopoietic Tissues

- In embryos, hemopoietic tissue is found distributed throughout the body

- In adults, hemopoiesis occurs primarily in the red bone marrow (which contains stem cells, the primordia of blood cells) and the spleen

Blood vessels are the first indicator of the formation of the circulatory system. Blood islands first form in the yolk and then become contiguous to form a network of vessels. Endothelium lines the blood vessels, and the rest is composed of muscular fibres, collagen and elastic fibres.

Blood Vessels have Three Layers of Tissues

1. The tunica intima is the inner layer of blood vessel that includes epithelium and elastic fibres
2. The tunica media is the middle layer of blood vessel that contains primarily smooth muscle fibres
3. The tunica externa is the extreme outer layer of the blood vessel that contains collagen fibres

Veins are usually larger in diameter and thinner-walled than arteries Veins channel blood to the heart, while arteries channel blood away from the heart Capillaries, the intermediaries between the arteries and the veins, are generally composed only of endothelium because they are where most of the diffusion in the circulatory system occurs — the junction at the capillaries marks an anastomosis, or a peripheral union between the blood vessels

The Heart

The embryonic heart is formed from the splanchnic layer of the mesoderm. *When first developed, the heart is composed of two layers*:

1. The endothelium forms the lining of the heart
2. The myocardium forms the muscular part of the heart, containing cardiac muscle fibres.

The primitive heart is a nearly straight tube having four parts which pumps a single stream of deoxygenated blood throughout the body:

- Blood flows from the sinus venosus, passes through the sinoatrial valve into the atrium, and then from the atrium into theventricle through the atrioventricular valve
- From the ventricle, the blood passes along a series of semilunar valves into the muscular conus arteriosus, and finally into the arterial system

Fishes

The heart of fishes differs very little from the ancestral vertebrate form:

- The heart is position very far forward in the body and lies adjacent to the gills
- Teleosts have lost the conus arteriosus and have developed the bulbus arteriosus, which is elastic and not muscular like the conus arteriosus

- Blood is moved through contraction of the respiratory hypobranchial muscles, which allows the sinus venosus to suck blood from the venous sinuses and propels blood into the ventricle

Amphibians and Reptiles

The amphibian heart is an intermediate three-chambered heart that allows for some separation of oxygenated and deoxygenated blood (Figure):

- The amphibian and reptile heart has two atria - one atrium receives blood returning from the body and one receives blood returning from the lungs
- The ventricle is undivided, so mixing of oxygenated and deoxygenated blood still can occur
- The only separation of blood is in the timing of when blood enters the ventricle
- After passing into the conus arteriosus blood flows into the truncus arteriosus, which bifurcates and travels through the rest of the body

Reptiles also have a three-chambered heart (Figure), but the circulation divides into three channels after the conus arteriosus — pulmonary trunk, right and left systemic trunks

Homeotherms

Homeotherms are characterized by having a four-chambered heart with a double circuit pump:

- The pulmonary circuit is located on the right side of the heart while the systemic circuit is found on the left side of the heart
- Homeotherms lack the sinus venosus that is found in more primitive hearts

Control of Heartbeat

The control of heartbeat in amniotes is influenced by the autonomic nervous system:

- The cardiac muscle itself has an inherent rhythm which assists in controlling the heartbeat.
- The sinoatrial node serves as the pacemaker of the heart and sets the initial rhythm of the heartbeat.
- The signal from the SA node is then conducted through the heart muscle by the Purkinje fibres, and the atrioventricular nodetransmits the signal through the cardiac muscle in the ventricle.

In lower vertebrates, the sinoatrial node is present, but the Purkinje fibres and the atrioventricular node are lacking

Coronary circulation is necessary to supply the metabolic needs of the cardiac muscle of mammals because it is larger in size than the two or three-chambered heart:

- The coronary circuit is composed of a pair of coronary arteries that leave the base of the aortic arch.
- Coronary veins return blood to the right atrium.

Arterial Channels and their Modifications

The arterial channels are vessels that comprise the initial functioning system of the embryo, and is basically the same for all vertebrates.

The heart:

- Pumps blood into the ventral aorta (truncus arteriosus)
- The ventral aorta then distributes blood into the aortic arches that run upward into the visceral arches
- Blood then enters the dorsal aorta from the aortic arches (Embryos of jawed animals usually have six aortic arches.)
- Blood from the anterior aortic arches runs forward into the head to the internal carotid arteries
- Blood from the posterior aortic arches runs into the dorsal aorta and posteriorly, and may branch into the vitelline or umbilical artery or any of the other intersegmental arteries

The first aortic arch (in the mandibular visceral arch) is always lost in the adult, along with the second aortic arch.

Aortic Arches of Fishes

Afferent branchial arteries lead into the gills from the aortic arches. Blood then flows through the gills through collector loops, and the oxygenated blood travels into the efferent branchial arteries, which continue into the dorsal aorta. Rostrally the dorsal aorta branches into the internal carotid arteries that supply oxygenated blood the head.

Caudally the dorsal aorta continues into the caudal artery, from which the following arteries offshoot:

- Coeliac and mesenteric (supply the abdominal viscera)
- Gonadal (supply the gonads)
- Renal (supply the kidneys)
- Intersegmental (associated with the myomeres)
- Subclavian (leads to the branchial arteries)
- Iliac (leads to the femoral arteries)

Aortic Arches of Tetrapods

Tetrapods also lack the first and second aortic arches.

The carotid system of tetrapods carries blood to the head and is derived from the third aortic arches.

It is composed of the common carotid arteries, which branch into:

- External carotid arteries, which supply the throat and the ventral part of the head
- Internal carotid arteries, which supply the brain and the rest of the head.

The left branch of the fourth aortic arch becomes the arch of the aorta in mammals, while the right one becomes the subclavian arteries. The sixth aortic arch becomes the pulmonary arteries, which are derived from a common pulmonary trunk on the aorta.

Posterior Arteries

The dorsal aorta is the large median longitudinal artery that extends posteriorly and eventually branches into the caudal artery.

Other branches of the dorsal aorta include:

- *Ventral visceral branches*: The celiac artery, which leads to the stomach, duodenum, liver and pancreas, and the mesenteric artery, which serves the remainder of the liver and the gut.
- *Lateral visceral branches*: These branches serve the urogenital organs (renal, ovarian, spermatic)
- *Dorsal somatic branches*: These branches serve the spinal cord, muscles and skin.

Venous Channels and their Modifications

The initial pattern of the venous channels comprises three systems:

1. Subintestinal-vitelline system
   - Drains the tail, digestive tract and yolk sac. Also includes the caudal vein, which runs to the cloacal area. The subintestinal veins continue drainage forward (after receiving blood from the vitelline veins) and eventually drains into the common cardinal vein.
2. Cardinal system
   - Drains the head, dorsal body wall and kidneys. Includes the anterior cardinal vein (lateral to the carotid arteries), posterior cardinal veins (lie adjacent to the kidneys) and the common cardinal veins (into which the posterior and anterior cardinal veins drain)
3. Abdominal system
   - Drains the ventral body wall and appendages. Consists of the lateral abdominal veins, which receive blood from the iliac and subclavian veins.

Anterior Veins

The anterior veins are derived from the cardinal system. In tetrapods, the anterior veins are composed by the internal and external jugular veins. The jugular veins unite with the subclavian vein and lead to the superior (cranial) vena cava, which also receives blood from the coronary veins (that first drain into the coronary sinus).

Hepatic Portal System

The hepatic portal system is derived from the subintestinal system. The hepatic portal vein receives blood from the gut region.

Inside the liver the vein breaks up into hepatic sinusoids, where the blood comes into contact with hepatic cells and phagocytic cells. Harmful materials are detoxified and removed from the blood in the liver.

Posterior to the liver the blood is collected into the hepatic vein, which joins with the caudal vena cava.

Renal Portal System

The renal portal system is derived from the posterior cardinal veins.

Blood from the posterior part of the body flows into the renal portal veins, which passes into the caudal vena cava.

The renal portal system is found only in fishes, amphibians, reptiles and birds. Thus, mammals have no renal portal system. All that remains in mammals is the azygous vein, which is an unpaired vein that drains most of the intercostal space on both sides of the mammalian thorax.

Posterior Veins

The posterior veins contain veins that are derived from one or all three initial systems (subintestinal, cardinal, abdominal). Adult birds and mammals lack the abdominal system, but as fetuses possessed two parts of the system (allantoic or umbilical veins).

Circulation in the Mammalian Fetus

Fetal mammals possess shunts between the pulmonary and systemic circuits because the placenta, not the lungs, is the site for gas exchange

Blood returns to the fetus from the placenta via the umbilical vein, enters the ductus venosus in the liver, and then passes into the caudal vena cava. The blood then passes through the foramen ovale (located in the septum between the right and left halves of the heart) and into the left atrium (thus bypassing the pulmonary circuit).

Since fetal lungs are not inflated, the pulmonary circuit is bypassed by the ductus arteriosus (a remnant of the sixth aortic arch) which joins to the aorta. At birth, lungs are inflated and the pulmonary circuit becomes more important in gas exchange. The pressure due to the flow of the blood from the lungs causes closure of the foramen ovale, leaving a grown-over region called the fossa ovalis. Due to lack of use as a shunt the ductus arteriosus closes and is filled with connective tissue to become the ligamentum arteriosum.

Lymphatic System

The purpose of the lymphatic system is to drain fluids that accumulate in the tissues and empty into the venous system.

Although Chondricthyes and other primitive fishes lack a true lymphatic system, they do have some vessels that help to drain the tissues (called the hemolymphatic system) that seems to be a precursor of the true lymphatic system. Components of the tetrapod lymphatic system include lymphatic capillaries to drain the tissues.

Across the vertebrate classes there is significant variation in the drainage of lymphatic capillaries into common larger ducts.

Amphibians and reptiles have three primary lymph vessels (subcutaneous, subvertebral, visceral) as well as lymph hearts, which are segmentally arranged masses containing smooth muscles that help propel lymph through the lymphatic system.

In birds and mammals the subvertebral ducts called thoracic ducts. Mammals also have the cisterna chyli, which is a sac that receives lymph from the abdominal viscera and caudal parts of the body. Mammals and birds also have lymph nodes, found in the neck, armpits and groin. Lymph nodes are the site of convergence for lymphatic vessels.

Other parts of the lymphatic system include the tonsils (lingual, pharyngeal and palatine), Peyer's patches (patches of lymphatic tissue in the small intestine of mammals), the vermiform appendix, and the bursa of Fabricius (a pouch in the cloaca of birds that contains lymphatic tissue. The thymus is a lymphatic organ that produces T lymphocytes, which are involved in the humoral immune response.

5

Vertebrate Circulatory System

CHAMBERS OF THE HEART

Vertebrate hearts contain muscular chambers called *atria* (sing. *atrium*) and ventricles. Contraction of the chamber forces blood out. Blood flows in one direction due to valves that prevent backflow.

The atrium functions to receive blood that is returning to the heart. When it contracts, blood is pumped into the *ventricle*.

The ventricle is the main pumping chamber of the heart. When it contracts, blood is pumped away from the heart to the body, lungs, or gills.

Circulatory System of Fish

In the diagrams that follow, arrows represent the direction of blood flow in blood vessels (arteries and veins). Blood pressure is represented by the thickness of the arrows. Thick arrows indicate high blood pressure. Blood that is rich in oxygen is represented by red arrows. Blue arrows represent blood that is low in oxygen after it has passed through the body tissues. Fish have a two-chambered heart with one atrium (A) and one ventricle (V).

The gills contain many capillaries for gas exchange, so the blood pressure is low after going through the gills. Low-pressure blood from the gills then goes directly to the body, which also has a large number of capillaries. The activity level of fish is limited due to the low rate of blood flow to the body.

Circulatory System of Amphibians

Amphibians have a 3-chambered heart with two atria and one ventricle.

Blood from the lungs (*pulmonary* flow) goes to the left atrium. Blood from the body (*systemic* flow) goes to the right atrium.

Both atria empty into the ventricle where some mixing occurs.

The advantage of this system is that there is high pressure in vessels that lead to both the lungs and body.

Circulatory System of Some Reptiles

In most reptiles, the ventricle is partially divided. This reduces mixing of oxygenated and unoxygenated blood in the ventricle. The partial division of the ventricle is represented by a dashed line below.

Circulatory System of Crocodilians, Birds, and Mammals

Birds and mammals (also crocodilians) have a four-chambered heart which acts as two separate pumps. After passing through the body, blood is pumped under high pressure to the lungs. Upon returning from the lungs, it is pumped under high pressure to the body. The high rate of oxygen-rich blood flow through the body enables birds and mammals to maintain high activity levels.

Blood Vessels

heart —> arteries —> arterioles —> capillaries —> venules —> veins —> heart

Arteries

Arteries carry blood away from heart.

Arteries have a thick, elastic layer to allow stretching and absorb pressure. The wall stretches and recoils in response to pumping, thus peaks in pressure are absorbed.

The arteries maintain pressure in the circulatory system much like a balloon maintains pressure on the air within it. The arteries therefore act as pressure reservoirs by maintaining (storing) pressure.

The elastic layer is surrounded by circular muscle to control the diameter and thus the rate of blood flow. An outer layer of connective tissue provides strength.

Arterioles

Smooth muscle surrounding the arteries and arterioles controls the distribution of blood. For example, blood vessels dilate when O_2 levels decrease or wastes accumulate. This allows more blood into an area to bring oxygen and nutrients or remove wastes.

Capillaries

The smallest blood vessels are capillaries. They are typically less than 1 mm long. The diameter is so small that red blood cells travel single file.

The total length of capillaries on one person is over 50,000 miles. This would go around the earth twice.

Not all of the capillary beds are open at one time because all of them would hold 1.4 times the total blood volume of the all the blood in the body.*Vasodilation* and *vasoconstriction* refer to the dilation and constriction of blood vessels. The diameter is controlled by neural and endocrine controls. Sphincter muscles control the flow of blood to the capillaries.

The total cross-sectional area of the capillaries is greater than that of the arteries or veins, so the rate of blood flow (velocity) is lowest in the capillaries. Blood pressure is highest in the arteries but is considerably reduced as it flows through the capillaries. It is lowest in the veins.

Interstitial fluid

The exchange of substances between blood and the body cells occurs in the capillaries. Capillaries are specialized for exchange of substances with the*interstitial fluid*. No cell in the body is more than 100 micrometers from a capillary. This is the thickness of four sheets of paper.

Interstitial fluid surrounds and bathes the cells. This fluid is continually being replaced by fresh fluid from blood in the circulatory system.

Body cells take up nutrients from the interstitial fluid and empty wastes into it. By maintaining a constant pH and ionic concentration of the blood, the pH and ionic concentration of the interstitial fluid is also stabilized.

Although fluid leaves and returns to the capillaries, blood cells and large proteins remain in the capillaries.

At the arterial end of capillaries, blood pressure forces fluid out and into the surrounding tissues. As blood moves through the capillary, the blood pressure decreases so that near the veinule end, less is leaking into the surrounding tissues.

As blood flows through the capillary and fluid moves out, the blood that remains behind becomes more concentrated. The osmotic pressure in the capillary is therefore greater near the veinule end and results in an increase in the amount of fluid moving into the capillary near this end.

The movement of blood into and out of the capillary. Long and thick arrows are used to represent a large amount of fluid movement. The total amount of movement out of the capillary is approximately equal to the amount of movement into the capillary. Notice that more blood tends to leave the capillary near the arteriole end and more tends to enter it near the veinule end.

The lymphatic capillaries collect excess fluid in the tissues.

Venules

Capillaries merge to form venules and venules merge into veins.

Venules can constrict due to the contraction of smooth muscle. When they are constricted there is more fluid loss in the capillaries due to increased pressure.

Veins

The diameter of veins is greater than that of arteries.

The blood pressure in the veins is low so valves in veins help prevent backflow.

The contraction of skeletal muscle during normal body movements squeezes the veins and assists with moving blood back to the heart.

The vena cava returns blood to the right atrium of the heart from the body. In the right atrium, the blood pressure is close to 0.

Varicose veins develop when the valves weaken.

Veins act as blood reservoirs because they contain 50% to 60% of the blood volume.

Smooth muscle in the walls of veins can expand or contract to adjust the flow volume returning to the heart and make more blood available when needed.

Portal Veins

Portal veins connect one capillary bed with another.

The hepatic portal vein connects capillary beds in the digestive tract with capillary beds in the liver.

HUMAN CIRCULATION

Chambers of the heart

The heart is actually two separate pumps. The left side pumps blood to the body (systemic circulation) and the right side pumps blood to the lungs (pulmonary circulation). Each side has an atrium and a ventricle.

The atria function to receive blood when they are relaxed and to fill the ventricles when they contract.

The ventricles function to pump blood to the body (left ventricle) or to the lungs (right ventricle).

Valves

Valves allow blood to flow through in one direction but not the other. They prevent backflow.

Atrioventricular valves are located between the atria and the ventricles. They are held in place by fibers called *chordae tendinae.*The left atrioventricular valve is often called the bicuspid or mitral valve; the right one is also called the tricuspid valve.

The *semilunar valves* are between the ventricles and the attached vessels.

The heartbeat sound is produced by the valves closing.

Below: The structure of the mammalian heart is summarized using a model.

Cardiac Cycle

As the *atria relax* and fill, the ventricles are also relaxed.

When the *atria contract,* the pressure forces the atrioventricular valves open and blood in the atria is pumped into the ventricles.

The *ventricles then contract,* forcing the atrioventricular valves closed. The pulmonary artery carries blood from the right ventricle to the lungs. The aorta carries blood from the left ventricle to the body.

Electrical stimulation

The heart does not require outside stimulation.

The sinoatrial (SA) node is a bit of nervous tissue that serves as the cardiac pacemaker. Stimulation from this node causes both of the atria to contract at the same time because the muscle tissue conducts the stimulation rapidly.

The contraction doesn't spread to the ventricles because the atria and ventricles are separated by connective tissue.

As a wave of stimulation (depolarization) spreads across the atria resulting in their contraction, another bit of nervous tissue called the atrioventricular (AV) node also becomes stimulated (depolarized). It conducts the action potential slowly to the ventricles. The slow speed is due to the small diameter of the neurons within the node.

The slow speed of conduction within the AV node ensures that the ventricles contract after the atria contract..

The bundle of His then transmits impulse rapidly from the AV node to the ventricles.

Nervous Control

Details of nervous control of the cardiac cycle are in the chapter on the nervous system.

Coronary circulation

Coronary arteries supply the heart muscles with blood.

They have a very small diameter and may become blocked, producing a heart attack.

Blood Pressure

The units of measurement are millimeters of mercury (mm Hg). For example, 120 mm Hg/80 mm Hg is considered to be normal blood pressure.

The top number is referred to as the systolic pressure; the bottom number is the diastolic pressure.

Hypertension - High Blood Pressure

High blood pressure is associated with cardiovascular disease.

In males under 45 years, pressures greater than 130/90 are considered to be high. In males over 45 years, pressures greater than 140 /95 are high.

Blood

Human blood has two parts, liquid (plasma) and cells.

Plasma

Plasma contains dissolved gasses, nutrients, wastes, salts, and proteins.

Salts and proteins buffer the pH so that it is approximately 7.4 and they maintain osmotic pressure.

Plasma proteins also assist in transporting large organic molecules. For example, lipoproteins carry cholesterol and albumin carries bilirubin (produced from the breakdown of hemoglobin when old blood cells are destroyed).

Cells

Red Blood Cells (Erythrocytes)

Red blood cells are biconcave disks filled with hemoglobin.

Red blood cells are continuously produced in the red marrow of the skull, ribs, vertebrae, and ends of the long bones. The nucleus of the cell disappears as it matures.

Mammalian red blood cells loose their nucleus as they mature. As a consequence, human red blood cells have a life span of approximately 120 days. Other vertebrates have nucleated red blood cells. Phagocytic cells in the liver and spleen remove old cells.

Anemia occurs when there are insufficient numbers of red blood cells or the cells lack sufficient hemoglobin.

White Blood Cells

White blood cells are covered in the chapter on the immune system.

Blood Clotting

Damaged tissue produces spasms of the smooth muscle and these spasms stop the blood flow for a few minutes. Platelets are fragments of larger cells produced in the bone marrow that assist in forming a clot. They adhere to exposed collagen in damaged blood vessels. This causes some to rupture and release substances that attract more platelets. Platelets and damaged tissue release substances that cause a blood protein called fibrinogen to be converted to fibrin. Fibrin forms a mesh-like structure that traps blood cells and platelets. The resulting plug that forms seals the leak.

Details of Blood Clot Formation

When tissue damage occurs, muscles begin to spasm, which temporarily reduces blood flow to the area. Blood flow is also reduced when platelets in the blood adhere to the damaged tissue. Blood clotting is initiated when platelets and damaged tissue secrete prothrombin activator.

The platelets and damaged tissue release a clotting factor called prothrombin activator.

Prothrombin activator and calcium ions catalyze the conversion of prothrombin to thrombin which then catalyzes the conversion of fibrinogen to fibrin threads.

Fibrin threads are sticky and trap more platelets, further sealing the leak.

BLOOD CIRCULATION IN VERTEBRATES

The heart is the central pump which drives the blood into the arteries. The arteries divide into smaller arterioles, ultimately becoming capillaries. The capillaries get transformed from arterial to the venous side where the venous trunk returns the blood back to the heart to complete the circulation. In its simplest form the heart must consist of two chambers in series.

Single Circulation

In fishes the heart is two chambered and S-shaped with four compartments in linear series. They are sinus venosus and atrium for receiving venous blood and a ventricle and conus arteriosus for pumping the blood. The blood passes through the heart to complete a circuit through gills and the body. The heart is a brachial heart as it pumps blood into the gill capillaries by the ventricle. The aorta is formed from these capillaries to split up again into the systemic capillaries in the body organs from where the blood is returned to the auricle. Thus there is only single circulation through the heart and it contains only venous blood.

Partial Double Circulation

In amphibians, Dipnoi fishes and some reptiles the heart is three chambered with two auricles and one ventricle. The two auricles are separated by inter auricular septum. The ventricle pumps the blood into the aorta which routes the major percentage of blood to the systemic vessels and some through the lungs. The oxygenated blood is brought back into the left auricle and the systemic deoxygenated blood into right auricle. The blood from both the auricles flow together into the ventricle. Hence the blood in the ventricle is a mixed blood and the circulation is termed as partial double circulation. Here the heart is described as pulmonary heart.

Complete Double Circulation

Beginning with the reptiles the ventricle begins to show a division into two parts by an inter-ventricular septum. This division is complete in crocodilians, birds and mammals which have two separate ventricles. This keeps the two blood streams separated as oxygenated blood in the left auricle and ventricle and deoxygenated blood in the right auricle and ventricle. A system of valves is present in heart which allows only one way conduction of blood. Thus the left ventricle pumps the oxygenated blood into the systemic vessels through aorta, and this blood in the venous state is brought back to the right auricle. From here blood flows into the right ventricle from where it is pumped into the lungs through pulmonary aorta. The oxygenated blood from lungs is brought back to the left ventricle through left auricle. As there is no mixing of arterial and venous blood at any stage it completes double circulation.

Flow of Blood

The pressure that develops within the closed vertebrate circulatory system is highest at the pump—the heart—and decreases with distance away from the pump because of friction within the blood vessels. Because the blood vessels can change their diameter, blood pressure can be affected by both the action of the heart and changes in the size of the peripheral blood vessels. Blood is a living fluid—it is viscous and contains cells (45 per cent of its volume in human beings)—and yet the effects of the cells on its flow patterns are small.

Blood enters the atrium by positive pressure from the venous system or by negative pressure drawing it in by suction. Both mechanisms operate in vertebrates. Muscular movements of the limbs and body, and gravity in land vertebrates, are forces propelling blood to the heart. In fishes and amphibians the atrium forces blood into the ventricle when it contracts. In birds and mammals the blood arrives at the heart with considerable residual pressure and passes through the auricles into the ventricles, apparently without much additional impetus from contraction of the auricles.

The ventricle is the main pumping chamber, but one of the features of double circulation is that the two circuits require different pressure levels. Although the shorter pulmonary circulation requires less pressure than the much longer systemic circuit, the two are connected to each other and must transport the same volume of fluid per unit time. The right and left ventricles in birds and mammals function as a volume and a pressure pump, respectively. The thick muscular wall of the left ventricle ensures that it develops a higher pressure during contraction in order to force blood through the body. It follows that pressures in the aorta and pulmonary artery may be very different. In human beings aortic pressure is about six times higher.

Valves throughout the system are crucial to maintain pressure. They prevent backflow at all levels; for example, they prevent flow from the arteries back into the heart as ventricular pressure drops at the end of a contraction cycle. Valves are important in veins, where the pressure is lower than in arteries.

Another impetus to blood flow is contraction of the muscles in the walls of vessels. This also prevents backflow of arterial blood towards the heart at the end of each contraction cycle. Input from nerves, sensory receptors in the vessels themselves, and hormones all influence blood vessel diameter, but responses differ according to position in the body and animal species. Normally, the pressures that develop in a circulatory system vary widely in different animals. Body size can be an important factor. The closed circulation systems of vertebrates generally operate at higher pressures than the open blood systems of invertebrates; the systems of birds and mammals operate at the highest pressures of all.

Electrical Activity

The vertebrate heart is myogenic (rhythmic contractions are an intrinsic property of the cardiac muscle cells themselves). Pulse rate varies widely in different vertebrates, but it is generally higher in small animals, at least in birds and mammals. Each chamber of the heart has its own contraction rate. In the frog, for example, the sinus venosus contracts fastest and is the pacemaker for the other chambers, which contract in sequence and at a decreasing rate, the conus being the slowest. In birds and mammals, where the sinus venosus is incorporated into the right atrium at the sinoauricular node, the latter is still the pacemaker and the heartbeat is initiated at that point. Thus, the evolutionary history of the heart explains the asymmetrical pattern of the heartbeat.

In the frog each contraction of the heart begins with a localized negative charge that spreads over the surface of the sinus venosus. In lower vertebrates, the cardiac muscle cells themselves conduct the wave of excitation. In birds and mammals, however, special conducting fibres (arising from modified muscle cells) transmit the wave of excitation from the sinoauricular node to the septum between the auricles, and then, after a slight delay, down between and around the ventricles. The electrical activity of the heart can be recorded; the resulting pattern is called an electrocardiogram.

Control of Heartbeat and Circulation

Many factors, such as temperature, oxygen supply, or nervous excitement, affect heartbeat and circulation. Blood circulation is controlled mainly via nerve connections, sensory receptors, and hormones. These act primarily by varying the heart's pulse rate, amplitude, or stroke volume and by altering the degree of dilation or constriction of the peripheral blood vessels (*i.e.*, those blood vessels near the surface of the body).

Temperature has a direct effect on heart rate, and one of the ways in which mammals regulate their internal temperature is by controlling peripheral blood circulation. Mammals are endothermic (warm-blooded) vertebrates; their internal temperature is kept within narrow limits by using heat generated by the body's own metabolic processes. Lizards are ectothermic (cold-blooded); they obtain heat from the external environment by, for example, basking in the sun. The effects of oxygen concentration on the heart and blood vessels is rapid. Oxygen deficiency in the cardiac tissue causes dilation of the coronary capillaries, thereby increasing blood flow and oxygen supply.

Most effects on the circulation are indirect and complex. All vertebrate hearts receive input from nerves; for example, stimulation of a branch of the vagus nerve causes the release of acetylcholine at the nerve endings, which depresses the heart rate. Other nerve endings release norepinephrine, which increases the heart rate. Less directly, nervous stimulation brought about by stress causes the release of the hormones epinephrine and norepinephrine into the bloodstream. These substances not only make the heart beat faster and with a greater amplitude, but they also divert blood to the muscles by constricting the vessels in the skin and gut. This prepares the animal physiologically for physical exertion. Numerous other chemicals, such as nicotine, affect heart rate directly or indirectly.

Two other factors are important in the context of circulatory regulation—the concentrations of inorganic ions and sensory receptors in blood vessel walls. Sodium, potassium, and calcium ions are always involved in changes of electrical potential across cell membranes. A change in their concentrations, therefore, influences heartbeat profoundly. External calcium concentration can, for example, determine the conductance of sodium across the cardiac cell membranes. Sensory receptors in the walls of blood vessels register blood pressure. They are found in the aorta, carotid arteries, pulmonary artery, capillaries in the adrenal gland, and the tissues of the heart itself. Impulses from the receptors travel to the medulla of the brain, from where messages are sent via motor nerves to the heart and blood vessels. Regulation is thus achieved according to the body's needs.

CIRCULATORY SYSTEM

Why don't we have cells as big as elephants?

Surface area to volume ratio decreases with size.This limits transport of nutrients and wastes.

Multicellular organisms have a problem getting materials to and from cells.

Many invertebrates have open circulatory systems where some important organs are bathed in blood and diffusion from other tissue allows exchange of nutrients and wastes.

All vertebrates have closed cardiovascular systems that move blood in an orderly and controlled manner around the body.

Although it is called a closed system in fact it is not completely closed. Fluid continually is lost from capillaries and enters the interstitial space between cells, carrying nutrients.

This fluid is returned to the cardiovascular system by the lymph system.

Closed systems are more efficient as it allows greater exchange rates because of the constant flow of blood.

The vertebrate body is approximately 70% water.

This is distributed in the body in three major categories

Intracellular water 45%

Interstitial water 20%

Plasma 5%

Cardiovascular system is responsible (together with the lymphatic system) for moving the water and maintaining homeostasis of the interstitial fluid that baths the cells. Cells are constantly exchanging ions, gases, nitrogenous waste, amino acids sugars etc. with the interstitial fluid in order to maintain there own internal environments.

Added to the distribution of solutes is the distribution of heat.

Each class of vertebrates has a uniform pattern of circulation but there are substantial differences between classes. A major difference in circulation pattern exists between water breathing fish and air breathers.

Fish

Fish have a two chambered heart with a single atrium and a single muscular ventricle.

Just before the atrium is an enlarged area called the sinus venosus that acts as a reservoir and receiving chamber for blood.

As blood leaves the ventricle in teleosts (bony fish) it passes through the bulbous arteriosis.

The bulbous arteriosus is an area of the aorta with thickened muscular walls which acts to absorb pressure and even blood flow from the ventricle.

In elasmobranchs blood leaving the ventricle passes through the conus arteriosus.

Valves are present on the conus arteriosus that prevent backflow of blood.

Blood flows forward from the heart towards the gills.

After passing through the gills and picking up oxygen it collects in arteries and is distributed to the body.

Amphibians

Along with air breathing amphibians have evolved a split circulation system, one systemic and one pulmocutaneous.

Amphibians have a three chambered heart.

Two atria and a single ventricle.

The left atria receives blood from the lungs and the right atria receives blood from the body.

As the blood passes from the atria to the single ventricle there is relatively little mixing of the blood.

A short conus arteriosus with a spiral valve directs blood as it leaves the ventricle.

Generally the blood from the left atria is directed toward the systemic artery (body) and the blood from the right atria is directed toward the pulmocutaneous artery.

The circulation via the pulmocutaneous artery is unique to amphibians.

It branches into the pulmonary and cutaneous arteries.

This allows the skin to also receive unoxygenated blood and facilitates diffusion across those surfaces.

At times when the lungs are not being used capillary recruitment in the skin provides increased blood to the skin and vasoconstriction decreased flow to the lungs.

With decreased flow from the lungs some of the blood returning from the skin moves to the systemic artery.

Reptiles

Lizards, snakes and turtles.

Lizards, snakes and turtles have a three chambered heart with two atria and one partially divided ventricle.

There is a well developed double circulation because under normal circumstances there is little mixing of blood in the ventricle.

These reptiles do have the ability to alter their circulation.

If pulmonary resistance is increased, systemic venous return will not all flow into pulmonary artery.

Some blood will be pumped primarily into the aorta on ventricular contraction.

This is termed the right-to-left shunt. i.e. the lungs are bypassed.

Right-to-left shunt is used by diving reptiles and by others during periods of apnea. Similarly, if resistance is low in the pulmonary circulation, a left-to-right shunt takes place.

Crocodilians

Crocodilians have a four chambered heart.

There are two aortic arches. The left aortic arch originates in the right ventricle and the right aortic arch originates in the left ventricle.

The two aortic arches are joined by the foramen of Panizzae.

The pulmonary artery originates in the right ventricle also.

Under normal(?) conditions, higher pressure from the left ventricle causes pressure in the left aortic arch to close a valve at its junction with the right ventricle and there is no mixing of oxygenated and non-oxygenated blood in the aorta.

During diving events however, muscular constriction of the pulmonary artery causes increased pressure in the right ventricle and blood passes from the right ventricle into the aorta and systemic circulation.

i.e. the lungs are bypassed - right to left shunt.

Shunting of blood has been reported to be as high as 100% in some diving reptiles.

Oxygen in lungs is used as a reservoir and intermittent changes in shunting has been observed.

It has also been suggested also as a mechanism for thermoregulation and to prevent oxygen loss through the skin.

Three chambered heart not imperfect version of mammal heart but organ that meets demands for ectotherms.

Birds and Mammals

Birds and mammals have a four chambered heart with separate pulmonary and systemic circulations.

Complete separation allows differences in blood pressure in these circulations.

Peripheral resistance is much lower in the pulmonary system.

Mean blood pressure in aorta of humans is approximately 100 mm Hg.

Mean blood pressure in pulmonary artery of humans is approximately 20 mm Hg.

RESPIRATORY SYSTEM IN VERTEBRATES

Cellular respiration involves the breakdown of organic molecules to produce ATP. A sufficient supply of oxygen is required for the aerobic respiratory machinery of Kreb's Cycle and the Electron Transport System to efficiently convert stored organic energy into energy trapped in ATP. Carbon dioxide is also generated by cellular metabolism and must be removed from the cell. There must be an exchange of gases: carbon dioxide leaving the cell, oxygen entering. Animals have organ systems involved in facilitating this exchange as well as the transport of gases to and from exchange areas.

Bodies and Respiration

Single-celled organisms exchange gases directly across their cell membrane. However, the slow diffusion rate of oxygen relative to carbon dioxide limits the size of single-celled organisms. Simple animals that lack specialized exchange surfaces have flattened, tubular, or thin shaped body plans, which are the most efficient for gas exchange. However, these simple animals are rather small in size.

Respiratory Surfaces

Large animals cannot maintain gas exchange by diffusion across their outer surface. They developed a variety of respiratory surfaces that all increase the surface area for exchange, thus allowing for larger bodies. A respiratory surface is covered with thin, moist epithelial cells that allow oxygen and carbon dioxide to exchange. Those gases can only cross cell membranes when they are dissolved in water or an aqueous solution, thus respiratory surfaces must be moist.

Methods of Respiration

Sponges and jellyfish lack specialized organs for gas exchange and take in gases directly from the surrounding water. Flatworms and annelids use their outer surfaces as gas exchange surfaces. Arthropods, annelids, and fish use gills; terrestrial vertebrates utilize internal lungs.

The Body Surface

Flatworms and annelids use their outer surfaces as gas exchange surfaces. Earthworms have a series of thin-walled blood vessels known as capillaries.

Gas exchange occurs at capillaries located throughout the body as well as those in the respiratory surface.

Amphibians use their skin as a respiratory surface. Frogs eliminate carbon dioxide 2.5 times as fast through their skin as they do through their lungs. Eels (a fish) obtain 60% of their oxygen through their skin. Humans exchange only 1% of their carbon dioxide through their skin. Constraints of water loss dictate that terrestrial animals must develop more efficient lungs.

Gills

Gills greatly increase the surface area for gas exchange. They occur in a variety of animal groups including arthropods (including some terrestrial crustaceans), annelids, fish, and amphibians. Gills typically are convoluted outgrowths containing blood vessels covered by a thin epithelial layer. Typically gills are organized into a series of plates and may be internal (as in crabs and fish) or external to the body (as in some amphibians).

Gills are very efficient at removing oxygen from water: there is only 1/20 the amount of oxygen present in water as in the same volume of air. Water flows over gills in one direction while blood flows in the opposite direction through gill capillaries. This countercurrent flow maximizes oxygen transfer.

Tracheal Systems

Many terrestrial animals have their respiratory surfaces inside the body and connected to the outside by a series of tubes.Tracheae are these tubes that carry air directly to cells for gas exchange. Spiracles are openings at the body surface that lead to tracheae that branch into smaller tubes known as tracheoles. Body movements or contractions speed up the rate of diffusion of gases from tracheae into body cells. However, tracheae will not function well in animals whose body is longer than 5 cm.

Lungs

Lungs are ingrowths of the body wall and connect to the outside by as series of tubes and small openings. Lung breathing probably evolved about 400 million years ago. Lungs are not entirely the sole property of vertebrates, some terrestrial snails have a gas exchange structures similar to those in frogs.

Principles

- Movement of an oxygen-containing medium so it contacts a moist membrane overlying blood vessels.
- Diffusion of oxygen from the medium into the blood.
- Transport of oxygen to the tissues and cells of the body.
- Diffusion of oxygen from the blood into cells.

- Carbon dioxide follows a reverse path.

The Human Respiratory System

This system includes the lungs, pathways connecting them to the outside environment, and structures in the chest involved with moving air in and out of the lungs.

Air enters the body through the nose, is warmed, filtered, and passed through the nasal cavity. Air passes the pharynx (which has the epiglottis that prevents food from entering the trachea).The upper part of the trachea contains the larynx. The vocal cords are two bands of tissue that extend across the opening of the larynx. After passing the larynx, the air moves into the bronchi that carry air in and out of the lungs.

Bronchi are reinforced to prevent their collapse and are lined with ciliated epithelium and mucus-producing cells. Bronchi branch into smaller and smaller tubes known as bronchioles. Bronchioles terminate in grape-like sac clusters known as alveoli. Alveoli are surrounded by a network of thin-walled capillaries. Only about 0.2 μm separate the alveoli from the capillaries due to the extremely thin walls of both structures.

The lungs are large, lobed, paired organs in the chest (also known as the thoracic cavity). Thin sheets of epithelium (pleura) separate the inside of the chest cavity from the outer surface of the lungs. The bottom of the thoracic cavity is formed by the diaphragm.

Ventilation is the mechanics of breathing in and out. When you inhale, muscles in the chest wall contract, lifting the ribs and pulling them, outward. The diaphragm at this time moves downward enlarging the chest cavity. Reduced air pressure in the lungs causes air to enter the lungs. Exhaling reverses theses steps.

Diseases of the Respiratory System

The condition of the airways and the pressure difference between the lungs and atmosphere are important factors in the flow of air in and out of lungs.

Many diseases affect the condition of the airways:

- Asthma narrows the airways by causing an allergy-induced spasms of surrounding muscles or by clogging the airways with mucus.
- Bronchitis is an inflammatory response that reduces airflow and is caused by long-term exposure to irritants such as cigarette smoke, air pollutants, or allergens.
- Cystic fibrosis is a genetic defect that causes excessive mucus production that clogs the airways.

The Alveoli and Gas Exchange

Diffusion is the movement of materials from a higher to a lower concentration. The differences between oxygen and carbon dioxide concentrations are measured by partial pressures. The greater the difference in partial pressure the greater the rate of diffusion.

Respiratory pigments increase the oxygen-carrying capacity of the blood. Humans have the red-colored pigment hemoglobin as their respiratory pigment. Hemoglobin increases the oxygen-carrying capacity of the blood between 65 and 70 times. Each red blood cell has about 250 million hemoglobin molecules, and each milliliter of blood contains 1.25×10^{15} hemoglobin molecules. Oxygen concentration in cells is low (when leaving the lungs blood is 97% saturated with oxygen), so oxygen diffuses from the blood to the cells when it reaches the capillaries. Carbon dioxide concentration in metabolically active cells is much greater than in capillaries, so carbon dioxide diffuses from the cells into the capillaries. Water in the blood combines with carbon dioxide to form bicarbonate. This removes the carbon dioxide from the blood so diffusion of even more carbon dioxide from the cells into the capillaries continues yet still manages to "package" the carbon dioxide for eventual passage out of the body.

In the alveoli capillaries, bicarbonate combines with a hydrogen ion (proton) to form carbonic acid, which breaks down into carbon dioxide and water. The carbon dioxide then diffuses into the alveoli and out of the body with the next exhalation.

Control of Respiration

Muscular contraction and relaxation controls the rate of expansion and constriction of the lungs. These muscles are stimulated by nerves that carry messages from the part of the brain that controls breathing, the medulla. Two systems control breathing: an automatic response and a voluntary response. Both are involved in holding your breath.

Although the automatic breathing regulation system allows you to breathe while you sleep, it sometimes malfunctions. Apnea involves stoppage of breathing for as long as 10 seconds, in some individuals as often as 300 times per night. This failure to respond to elevated blood levels of carbon dioxide may result from viral infections of the brain, tumors, or it may develop spontaneously. A malfunction of the breathing centers in newborns may result in SIDS (sudden infant death syndrome). As altitude increases, atmospheric pressure decreases. Above 10,000 feet decreased oxygen pressures causes loading of oxygen into hemoglobin to drop off, leading to lowered oxygen levels in the blood. The result can be mountain sickness (nausea and loss of appetite). Mountain sickness does not result from oxygen starvation but rather from the loss of carbon dioxide due to increased breathing in order to obtain more oxygen.

CHARACTERS OF RESPIRATORY TISSUE IN VERTEBRATES

Oxygen is needed by aerobic organisms because it is the final electron acceptor during cellular respiration. The diagram below shows that Cellular respiration is a process in which electrons are removed from glucose in a series of steps. The electrons are carried by NADH and FADH2 to the electron transport system. The electron transport system uses the energy in the electrons to synthesize ATP. The remaining carbon atoms in the glucose molecule are released as CO_2, a waste product. The equation for the complete breakdown of glucose by aerobic eukaryotes is:

$$C_6H_{12}O_6 + 6O_2 \rightarrow 6CO_2 + 6H_2O + 36\ ATP$$

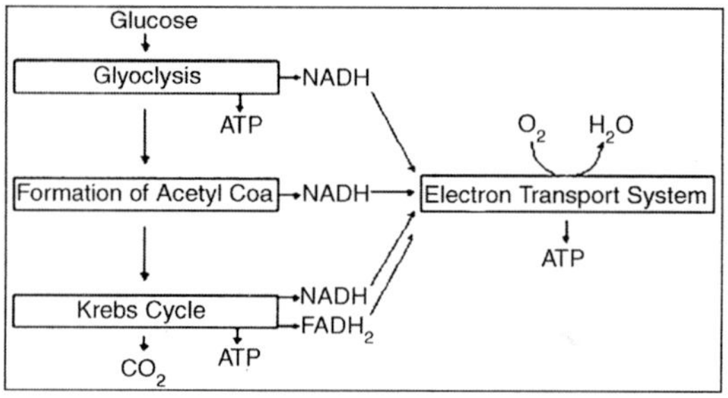

Atmosphere

78% N_2, 21% O_2, 1% argon, noble gases, CO_2

Properties

Diffusion refers to movement of molecules from an area of higher concentration to an area of lower concentration.

Partial pressure is the pressure exerted by one gas in a mixture.

Total atmospheric pressure at sea level = 760 mm Hg.

Partial pressure O_2 = 760 × .21 = 160 mm Hg.

Gasses move by diffusion from areas of higher partial pressure to areas of lower partial pressure.

Respiratory Surfaces

All animals need to take in O_2 and eliminate CO_2. *Lungs* are membranous structures designed for gas exchange in a terrestrial environment. *Gills* are designed for gas exchange in an aquatic environment. Oxygen must be dissolved in water before animals can take it up. Therefore, the respiratory surfaces of animals (gills, lungs, etc.) must always be moist. This is true of all animals. Very small organisms don't need respiratory surfaces because they have a high surface:volume ratio.

Skin

The skin can be used as a respiratory surface but it does not have much surface area compared to lungs or gills. Animals that rely on their skin as a respiratory organ are small and either have low metabolic rates or they also have lungs or gills.

Like all respiratory surfaces, the skin must remain moist to function in gas exchange. Amphibians, most annelids, some mollusks, and some arthropods use their skin as a respiratory organ.

Gills

Gills provide a large surface area for gas exchange in aquatic organisms. It is difficult to circulate water past gills because water is dense and the O_2 concentration in water is low. There is 5% as much oxygen in water as there is in air. To circulate water past the gills, amphibian larvae physically move their gills, mollusks pump water into mantle cavity which contains the gills, and some crustacean gills are attached to branches of the walking legs. The flow of blood in the gills of fish is in the opposite direction that water passes over the gills. This arrangement (called *countercurrent flow*) enables fish to extract more oxygen from the water than if blood moved in the same direction as the passing water. Gills cannot be used in air because they lack structural support; they would collapse. Their use in air would also result in too much water loss by evaporation.

Tracheal System

Insects, centipedes, and some mites and spiders have a tracheal respiratory system.

Tracheae are a network of tubules that bring oxygen directly to the tissues and allow carbon dioxide to escape. The openings to the outside, called *spiracles*, are located on the side of the abdomen.

Trachea and lungs are internal to reduce water loss.

Vertebrate Lungs

- Simple lungs evolved 450 million years ago in fish.
- Some evolved into swim bladders.
- Others evolved into more complex lungs.
- Paired lungs are the respiratory surfaces in all reptiles, birds, and mammals.

Amphibians

lung is a simple convoluted sac have small lungs but obtain much O_2 by diffusion across moist skin ventilate lungs by *positive pressure*; (reptiles, birds and mammals use negative pressure)

Reptiles

The skin is watertight; it is not used as a respiratory surface. The lungs possess alveoli. All diffusion occurs across the alveolar surface.

Birds and Mammals

The lungs of birds and mammals are more branched with smaller, more numerous alveoli.

Birds

Birds have one-way flow of air in their lungs. As a result, the lungs receive fresh air during inhalation and again during exhalation.

Advantages of one-way flow:

- No residual volume; all old (stale) air leaves with each breath
- Crosscurrent flow (crosscurrent = 90°; countercurrent = 180°; crosscurrent is not as efficient but is still more efficient than mammalian lung)

One-way flow is accomplished by the use of air sacs as illustrated below. During inspiration, the air sacs fill. During expiration, they empty.

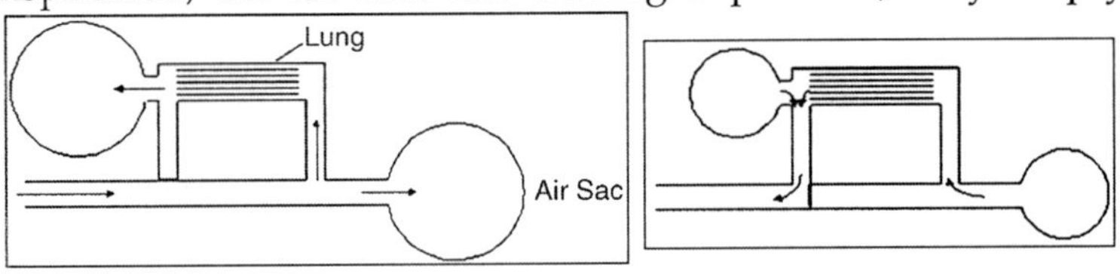

Human Respiratory System

Surface area of human lung is 60 to 80 sq. meters

Structures

Pharynx → epiglottis (open space is the glottis) → larynx with vocal cords → trachea → bronchi → bronchioles → alveoli

Nasal Cavities

- Hair and cilia filter dust and particles.
- Blood vessels warm air and mucus moistens air.

Ventilation

- To inhale, the diaphragm contracts and flattens.
- Muscles move the rib cage which also contributes to expanding the chest cavity.

To exhale, the muscles relax and elastic lung tissue recoils.

The Heimlich Maneuver

Choking results when food enters the trachea instead of the esophagus. The Heimlich maneuver can force air out of the lungs to dislodge the obstruction.

Respiratory Pigments

Hemoglobin

Hemoglobin is a protein that carries oxygen and is found in the blood of most animals. It is synthesized by and is contained within erythrocytes (red blood cells). Oxygen is bound reversibly to the iron portion. Hemoglobin increases the oxygen-carry capacity of the blood by 70 times. 95% of the oxygen is transported by hemoglobin, 5% in blood plasma.

The bright red colour occurs when it is bound with oxygen.

Hemocyanin

Hemocyanin is a carrier protein found in many invertebrates:

- It uses copper instead of iron.
- It does not occur within blood cells; it exists free in the blood. (Their blood is called hemolymph.)
- It is bright blue when bound with oxygen.

Gas Exchange and Transport

Gas Exchange in humans occurs in alveoli. Gasses must diffuse across the alveolar wall, a thin film of interstitial fluid, and the capillary wall.

Partial Pressures

	Lungs	*Tissues*
Oxygen	***High***	***Low***
CO_2	***Low***	***High***

The partial pressure of CO_2 is higher in the tissues because respiring tissues produce CO_2 as a result of the breakdown of glucose ($C_6H_{12}O_6$) during cellular respiration.

Oxygen Transport

Hemoglobin molecule + 4 oxygen molecules → oxyhemoglobin.

The amount of oxygen that combines depends upon the partial pressure. More oxygen is loaded at higher partial pressures of oxygen.

Hemoglobin does not necessarily release (unload) all of its oxygen as it passes through the body tissues.

Oxyhemoglobin releases its oxygen when:

- The partial pressure of O_2 is low.
- The partial pressure of CO_2 is high. High CO_2 causes the shape of the hemoglobin molecule to change and this augments the unloading of oxygen.
- The temperature is high.
- The pH is low (high acidity).

Active tissues need more oxygen and all of the conditions listed above are characteristic of actively metabolizing tissues. Therefore, these tissues receive more oxygen from hemoglobin than less active tissues.

CO (carbon monoxide) binds to hemoglobin 200 times faster than O_2 and does not readily dissociate from the hemoglobin. Small amounts of CO can cause respiratory failure.

Carbon Dioxide Transport

Carbon dioxide is transported to the lungs by one of the following ways:

- Dissolved CO_2
- Bound to hemoglobin ($HbCO_2$)
- HCO_3^- (bicarbonate ions).

Most is transported as bicarbonate ions because...

$$CO_2 + H_2O \leftrightarrow H_2CO_3 \leftrightarrow HCO_3^- + H^+$$

Low ← CO_2 partial pressure → High

(lungs) (tissues)

The equation above moves towards the right when the partial pressure of CO_2 is high. When the partial pressure of CO_2 is low, it moves to the left and CO_2 comes out of solution.

In the active tissues, the CO_2 partial pressure is high, so CO_2 becomes dissolved in water, forming H_2CO_3, which then forms HCO_3^- and H^+. In the lungs, the partial pressure of CO_2 is low because the concentration of CO_2 in the atmosphere is low. As blood passes through the lungs, HCO_3^- + H^+ form H_2CO_3 which then forms CO_2 + H_2O.

Carbonic anhydrase (in red blood cells) speeds up this reaction 150 times. HCO_3^- tends to diffuse out of the red blood cells into the plasma.

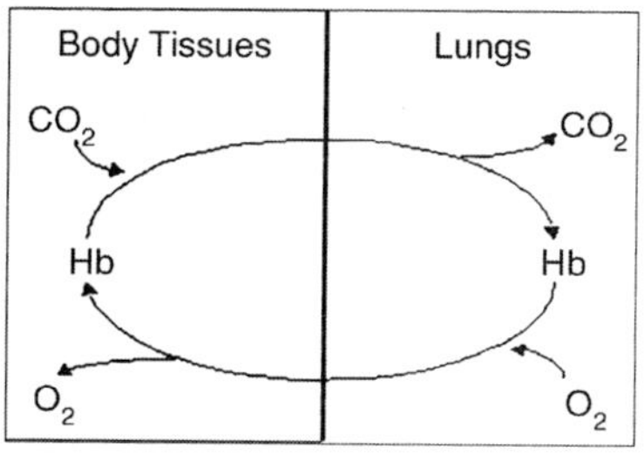

Control of Breathing Rate

Eliminating CO_2 is usually a bigger problem for terrestrial vertebrates than obtaining O_2. The body is therefore more sensitive to high CO_2 concentration than low O_2 concentration.

Aquatic vertebrates are more sensitive to low O_2 because O_2 is more limited in aquatic environments.

Neural Control Mechanisms in Terrestrial Vertebrates

During inhalation, the diaphragm and intercostal muscles are stimulated. Other neurons inhibit these when exhaling. Respiration is not under voluntary control.

Monitoring H^+ and CO_2

Chemoreceptors in the respiratory control center of the brain (medulla oblongata) detect changes in CO_2 by monitoring pH of cerebrospinal fluid. High CO_2 lowers the pH (an acid is a solution with a high H^+ concentration).

$$CO_2 + H_2O \leftrightarrow H_2CO_3 \leftrightarrow HCO_3^- + H^+$$

Chemoreceptors in the aorta and carotid artery are also sensitive to pH and to greatly reduced amounts of O_2.

Bronchiole Diameter

The primary bronchus branches extensively into bronchioles. Terminal bronchioles are surrounded by smooth muscle.

The diameter of the bronchioles (and blood vessels) increases or decreases in response to needs. It is adjusted by smooth muscle under the control of the nervous system. The parasympathetic nervous system stimulates these muscles to contract, reducing the diameter of the airways. This is advantageous when the body is relaxing and breathing is shallow. Narrow bronchioles result in less air remaining within the lungs after each exhalation. The sympathetic nervous system relaxes these muscles as a response to stressful situations. This allows a more rapid rate of intake and expulsion of air. Allergens trigger histamine release which constricts muscles. Narrower bronchioles result in decreased ventilation of the lungs. Severe attacks may be life-threatening.

Defence Mechanisms in the Respiratory Tract

Large particles are filtered out by the nose.

Small particles become trapped in mucous and are moved upward, out of the respiratory tract by cilia lining the bronchi and bronchioles.

Bronchitis

Bronchitis is an inflammation of the airways that causes mucous to accumulate. The normal cleansing activity of cilia is reduced and not sufficient to remove the mucous. Coughing attempts to clear the mucus. Smoking and other irritants increase mucus secretion and diminish cilia function.

Emphysema

- Emphysema occurs when the alveolar walls lose their elasticity. Damage to the walls also reduces the amount of surface available for gas exchange.
- Emphysema is associated with environmental conditions, diet, infections, and genetics. It can result from chronic bronchitis when the airways become clogged with mucous and air becomes trapped within the alveoli.

TRENDS IN ORGAN SYSTEMS - VERTEBRATE CIRCULATORY SYSTEMS

The metabolic activity of any tissue is limited by its blood supply; the more active any organ, the more blood it needs and the more extensive its vascularization.

The changes in metabolic activity associated with endothermy and the change from gill to lung respiration has led to changes in vertebrate circulatory systems.

Circulatory Systems

In general, blood is pumped by heart to arteries -> arterioles -> capillaries.

Capillaries come together to form venioles -> veins -> major venous trunks

-> heart

Veins between two capillary networks are portal systems.

Closed circulatory system, but fluid constituents of blood leak out of capillaries and return to the heart by the second component of the circulatory system, the lymphatic system.

Considerable adaptability built into vertebrate circulatory systems:

Considerable variation in blood vessels, as seen in shark

Due to developmental origin of blood vessels: many channels form, some enlarge to form major blood vessels, some of which atrophy and die.

A piece of a vein grafted into an artery transforms structurally to become an artery. Nearly any vessel can be tied off gradually and system can enlarge alternate routes.

Blood can, and does, flow in either direction in several vessels. Blood can be diverted towards or away from any part of the body, volume of circulating blood can be changed, rate of circulation can be changed 5-fold.

Despite this variability, the structure of the heart and major vessels tells us a lot about the evolutionary history of vertebrates and the evolutionary changes that have greatly modified vertebrate circulatory systems.

Circulatory system is the first organ system to become functional during development.

Development of Heart

On each side of embryo, an aorta forms ventrally and swings dorsally.

The paired aortae in ventral position fuse along a short section and form a single vessel, the beginning of the heart.

The posterior part of the fusion forms the sinus venosus, receiving the major venous trunks flowing back to the heart.

Four sequential chambers of the heart exist, from posterior to anterior:

1. thin-walled sinus venosus
2. single atrium
3. thick walled single ventricle
4. conus arteriosus

The adult heart (fish or amniote) develops from this simple pattern.

The heart takes on its final adult form from a number of possible modifications:

differential growth of certain parts

fusion of adjacent regions

disappearance of partitions

formation of new partitions

In amniotes, the heart overgrows its pericardial space and doubles up on itself to form a loop that swings to the right side of the pericardial cavity.

This twisting results in the shape of the heart and the positioning of the atria and ventricles seen in adult amniotes.

TYPES OF CIRCULATORY SYSTEMS

Living things must be capable of transporting nutrients, wastes and gases to and from cells. Single-celled organisms use their cell surface as a point of exchange with the outside environment. Multicellular organisms have developed transport and circulatory systems to deliver oxygen and food to cells and remove carbon dioxide and metabolic wastes. Sponges are the simplest animals, yet even they have a transport system. Seawater is the medium of transport and is propelled in and out of the sponge by ciliary action. Simple animals, such as the hydra and planaria, lack specialized organs such as hearts and blood vessels, instead using their skin as an exchange point for materials. This, however, limits the size an animal can attain. To become larger, they need specialized organs and organ systems.

Multicellular animals do not have most of their cells in contact with the external environment and so have developed circulatory systems to transport nutrients, oxygen, carbon dioxide and metabolic wastes. Components of the circulatory system include

- blood: a connective tissue of liquid plasma and cells
- heart: a muscular pump to move the blood
- blood vessels: arteries, capillaries and veins that deliver blood to all tissues

There are several types of circulatory systems. The open circulatory system, examples of which are diagrammed is common to molluscs and arthropods. Open circulatory systems (evolved in insects, mollusks and other invertebrates) pump blood into a hemocoel with the blood diffusing back to the circulatory system between cells. Blood is pumped by a heart into the body cavities, where tissues are surrounded by the blood. The resulting blood flow is sluggish.

Vertebrates, and a few invertebrates, have a closed circulatory system. Closed circulatory systems (evolved in echinoderms and vertebrates) have the blood closed at all times within vessels of different size and wall thickness. In this type of system, blood is pumped by a heart through vessels, and does not normally fill body cavities. Blood flow is not sluggish. Hemoglobin causes vertebrate blood to turn red in the presence of oxygen; but more importantly hemoglobin molecules in blood cells transport oxygen. The human closed circulatory system is sometimes called the cardiovascular system. A secondary circulatory system, the lymphatic circulation, collects fluid and cells and returns them to the cardiovascular system.

Vertebrate Cardiovascular System

The vertebrate cardiovascular system includes a heart, which is a muscular pump that contracts to propel blood out to the body through arteries, and a series of blood vessels. The upper chamber of the heart, the atrium (pl. atria), is where the blood enters the heart. Passing through a valve, blood enters the lower chamber, the ventricle. Contraction of the ventricle forces blood from the heart through an artery. The heart muscle is composed of cardiac muscle cells.

Arteries are blood vessels that carry blood away from heart. Arterial walls are able to expand and contract. Arteries have three layers of thick walls. Smooth muscle fibers contract, another layer of connective tissue is quite elastic, allowing the arteries to carry blood under high pressure.

The aorta is the main artery leaving the heart. The pulmonary artery is the only artery that carries oxygen-poor blood. The pulmonary artery carries deoxygenated blood to the lungs. In the lungs, gas exchange occurs, carbon dioxide diffuses out, oxygen diffuses in.Arterioles are small arteries that connect larger arteries with capillaries. Small arterioles branch into collections of capillaries known as capillary beds.

The thin-walled blood vessels in which gas exchange occurs. In the capillary, the wall is only one cell layer thick. Capillaries are concentrated into capillary beds. Some capillaries have small pores between the cells of the capillary wall, allowing materials to flow in and out of capillaries as well as the passage of white blood cells. Changes in blood pressure also occur in the various vessels of the circulatory system. Nutrients, wastes, and hormones are exchanged across the thin walls of capillaries. Capillaries are microscopic in size, although blushing is one manifestation of blood flow into capillaries. Control of blood flow into capillary beds is done by nerve-controlled sphincters.

The circulatory system functions in the delivery of oxygen, nutrient molecules, and hormones and the removal of carbon dioxide, ammonia and other metabolic wastes. Capillaries are the points of exchange between the blood and surrounding tissues. Materials cross in and out of the capillaries by passing through or between the cells that line the capillary.

The extensive network of capillaries in the human body is estimated at between 50,000 and 60,000 miles long. Thoroughfare channels allow blood to bypass a capillary bed. These channels can open and close by the action of muscles that control blood flow through the channels.

Blood leaving the capillary beds flows into a progressively larger series of venules that in turn join to form veins. Veins carry blood from capillaries to the heart. With the exception of the pulmonary veins, blood in veins is oxygen-poor. The pulmonary veins carry oxygenated blood from lungs back to the heart. Venules are smaller veins that gather blood from capillary beds into veins. Pressure in veins is low, so veins depend on nearby muscular contractions to move blood along. The veins have valves that prevent back-flow of blood.

Ventricular contraction propels blood into arteries under great pressure. Blood pressure is measured in mm of mercury; healthy young adults should have pressure of ventricular systole of 120mm, and 80 mm at ventricular diastole. Higher pressures (human 120/80 as compared to a 12/1 in lobsters) mean the volume of blood circulates faster (20 seconds in humans, 8 minutes in lobsters).

As blood gets farther from the heart, the pressure likewise decreases. Each contraction of the ventricles sends pressure through the arteries. Elasticity of lungs helps keep pulmonary pressures low.

Systemic pressure is sensed by receptors in the arteries and atria. Nerve messages from these sensors communicate conditions to themedulla in the brain. Signals from the medulla regulate blood pressure.

VERTEBRATE VASCULAR SYSTEMS

Humans, birds, and mammals have a four-chambered heart that completely separates oxygen-rich and oxygen-depleted blood. Fish have a two-chambered heart in which a single-loop circulatory pattern takes blood from the heart to the gills and then to the body. Amphibians have a three-chambered heart with two atria and one ventricle. A loop from the heart goes to the pulmonary capillary beds, where gas exchange occurs. Blood then is returned to the heart. Blood exiting the ventricle is diverted, some to the pulmonary circuit, some to systemic circuit. The disadvantage of the three-chambered heart is the mixing of oxygenated and deoxygenated blood. Some reptiles have partial separation of the ventricle. Other reptiles, plus, all birds and mammals, have a four-chambered heart, with complete separation of both systemic and pulmonary circuits.

The Heart

The heart, is a muscular structure that contracts in a rhythmic pattern to pump blood. Hearts have a variety of forms: chambered hearts in mollusks and vertebrates, tubular hearts of arthropods, and aortic arches of annelids. Accessory hearts are used by insects to boost or supplement the main heart's actions. Fish, reptiles, and amphibians have lymph hearts that help pump lymph back into veins.

The basic vertebrate heart, such as occurs in fish, has two chambers. An auricle is the chamber of the heart where blood is received from the body. A ventricle pumps the blood it gets through a valve from the auricle out to the gills through an artery.

Amphibians have a three-chambered heart: two atria emptying into a single common ventricle. Some species have a partial separation of the ventricle to reduce the mixing of oxygenated (coming back from the lungs) and deoxygenated blood (coming in from the body). Two sided or two chambered hearts permit pumping at higher pressures and the addition of the pulmonary loop permits blood to go to the lungs at lower pressure yet still go to the systemic loop at higher pressures.

Establishment of the four-chambered heart, along with the pulmonary and systemic circuits, completely separates oxygenated from deoxygenated blood. This allows higher the metabolic rates needed by warm-blooded birds and mammals.

The human heart is a two-sided, four-chambered structure with muscular walls. An atrioventricular (AV) valveseparates each auricle from ventricle. A semilunar (also known as arterial) valve separates each ventricle from its connecting artery.

The heart beats or contracts approximately 70 times per minute. The human heart will undergo over 3 billion contraction cycles, during a normal lifetime. The cardiac cycle consists of two parts: systole (contraction of the heart muscle) and diastole(relaxation of the heart muscle). Atria contract while ventricles relax. The pulse is a wave of contraction transmitted along the arteries. Valves in the heart open and close during the cardiac cycle. Heart muscle contraction is due to the presence of nodal tissue in two regions of the heart. The SA node (sinoatrial node) initiates heartbeat. The AV node (atrioventricular node) causes ventricles to contract. The AV node is sometimes called the pacemaker since it keeps heartbeat regular. Heartbeat is also controlled by nerve messages originating from the autonomic nervous system.

Blood flows through the heart from veins to atria to ventricles out by arteries. Heart valves limit flow to a single direction. One heartbeat, or cardiac cycle, includes atrial contraction and relaxation, ventricular contraction and relaxation, and a short pause. Normal cardiac cycles (at rest) take 0.8 seconds. Blood from the body flows into the vena cava, which empties into the right atrium. At the same time, oxygenated blood from the lungs flows from the pulmonary vein into the left atrium. The muscles of both atria contract, forcing blood downward through each AV valve into each ventricle.

Diastole is the filling of the ventricles with blood. Ventricular systole opens the SL valves, forcing blood out of the ventricles through the pulmonary artery or aorta. The sound of the heart contracting and the valves opening and closing produces a characteristic "lub-dub" sound. Lub is associated with closure of the AV valves, dub is the closing of the SL valves.

Human heartbeats originate from the sinoatrial node (SA node) near the right atrium. Modified muscle cells contract, sending a signal to other muscle cells in the heart to contract. The signal spreads to the atrioventricular node (AV node). Signals carried from the AV node, slightly delayed, through bundle of His fibers and Purkinjie fibers cause the ventricles to contract simultaneously.

Heartbeats are coordinated contractions of heart cardiac cells. When two or more of such cells are in proximity to each other their contractions synch up and they beat as one.

An electrocardiogram (ECG) measures changes in electrical potential across the heart, and can detect the contraction pulses that pass over the surface of the heart.

There are three slow, negative changes, known as P, R, and T. Positive deflections are the Q and S waves. The P wave represents the contraction impulse of the atria, the T wave the ventricular contraction. ECGs are useful in diagnosing heart abnormalities.

DISEASES OF THE HEART AND CARDIOVASCULAR SYSTEM

Cardiac muscle cells are serviced by a system of coronary arteries. During exercise the flow through these arteries is up to five times normal flow. Blocked flow in coronary arteries can result in death of heart muscle, leading to a heart attack.

Blockage of coronary arteries is usually the result of gradual buildup of lipids and cholesterol in the inner wall of the coronary artery.

Occasional chest pain, angina pectoralis, can result during periods of stress or physical exertion.

Angina indicates oxygen demands are greater than capacity to deliver it and that a heart attack may occur in the future. Heart muscle cells that die are not replaced since heart muscle cells do not divide. Heart disease and coronary artery disease are the leading causes of death in the United States.

Hypertension, high blood pressure (the silent killer), occurs when blood pressure is consistently above 140/90. Causes in most cases are unknown, although stress, obesity, high salt intake, and smoking can add to a genetic predisposition. Luckily, when diagnosed, the condition is usually treatable with medicines and diet/exercise.

THE VASCULAR SYSTEM

Two main routes for circulation are the pulmonary (to and from the lungs) and the systemic (to and from the body). Pulmonary arteries carry blood from the heart to the lungs. In the lungs gas exchange occurs. Pulmonary veins carry blood from lungs to heart. The aorta is the main artery of systemic circuit. The vena cavae are the main veins of the systemic circuit. Coronary arteries deliver oxygenated blood, food, etc. to the heart. Animals often have a portal system, which begins and ends in capillaries, such as between the digestive tract and the liver.

Fish pump blood from the heart to their gills, where gas exchange occurs, and then on to the rest of the body. Mammals pump blood to the lungs for gas exchange, then back to the heart for pumping out to the systemic circulation. Blood flows in only one direction.

Blood

Plasma is the liquid component of the blood. Mammalian blood consists of a liquid (plasma) and a number of cellular and cell fragment components. Plasma is about 60 % of a volume of blood; cells and fragments are 40%. Plasma has 90% water and 10% dissolved materials including proteins, glucose, ions, hormones, and gases. It acts as a buffer, maintaining pH near 7.4. Plasma contains nutrients, wastes, salts, proteins, etc. Proteins in the blood aid in transport of large molecules such as cholesterol.

Red blood cells, also known as erythrocytes, are flattened, doubly concave cells about 7 μm in diameter that carry oxygen associated in the cell's hemoglobin. Mature erythrocytes lack a nucleus. They are small, 4 to 6 million cells per cubic millimeter of blood, and have 200 million hemoglobin molecules per cell. Humans have a total of 25 trillion red blood cells (about 1/3 of all the cells in the body). Red blood cells are continuously manufactured in red marrow of long bones, ribs, skull, and vertebrae. Life-span of an erythrocyte is only 120 days, after which they are destroyed in liver and spleen. Iron from hemoglobin is recovered and reused by red marrow. The liver degrades the heme units and secretes them as pigment in the bile, responsible for the color of feces. Each second two million red blood cells are produced to replace those thus taken out of circulation. White blood cells, also known as leukocytes, are larger than erythrocytes, have a nucleus, and lack hemoglobin. They function in the cellular immune response. White blood cells (leukocytes) are less than 1% of the blood's volume. They are made from stem cells in bone marrow. There are five types of leukocytes, important components of the immune system. Neutrophils enter the tissue fluid by squeezing through capillary walls and phagocytozing foreign substances. Macrophages release white blood cell growth factors, causing a population increase for white blood cells. Lymphocytes fight infection. T-cells attack cells containing viruses. B-cells produce antibodies. Antigen-antibody complexes are phagocytized by a macrophage. White blood cells can squeeze through pores in the capillaries and fight infectious diseases in interstitial areas

Platelets result from cell fragmentation and are involved with clotting. Platelets are cell fragments that bud off megakaryocytes in bone marrow. They carry chemicals essential to blood clotting. Platelets survive for 10 days before being removed by the liver and spleen. There are 150,000 to 300,000 platelets in each milliliter of blood. Platelets stick and adhere to tears in blood vessels;

they also release clotting factors. A hemophiliac's blood cannot clot. Providing correct proteins (clotting factors) has been a common method of treating hemophiliacs. It has also led to HIV transmission due to the use of transfusions and use of contaminated blood products.

The Lymphatic System

Water and plasma are forced from the capillaries into intracellular spaces. This interstitial fluid transports materials between cells. Most of this fluid is collected in the capillaries of a secondary circulatory system, the lymphatic system. Fluid in this system is known as lymph.

Lymph flows from small lymph capillaries into lymph vessels that are similar to veins in having valves that prevent backflow. Lymph vessels connect to lymph nodes, lymph organs, or to the cardiovascular system at the thoracic duct and right lymphatic duct.

Lymph nodes are small irregularly shaped masses through which lymph vessels flow. Clusters of nodes occur in the armpits, groin, and neck. Cells of the immune system line channels through the nodes and attack bacteria and viruses traveling in the lymph.

6

Mammal Vertebrate

MAMMALIAN CHARACTERISTICS

Many of the most important and diagnostic mammalian characteristics serve to further intelligence and sensibility, promote endothermy, or to increase the efficiency of reproduction or the securing and processing of food. Basic structural body plan is inherited from Therapsid mammal-like reptiles.

Survival through mammalian evolution was perhaps due to their ability to move and to think more quickly than their Archosaurian counterparts.

Morphological trends were toward structural simplification:

- skull and jaw bones lost or reduced in size
- limbs and limb girdles simplified, reduced, and less laterally splayed

Fossil record provides little evidence on when endothermy actually developed.Diagnostic or Distinguishable Characteristics of Mammals:

Soft Tissues

Skin glands: Mammalian skin contains several kinds of glands not found in other vertebrates.

Mammary Glands: Provide nourishment for the young during their postnatal period of rapid growth.

Milk Composition: Milk composition varies with species:

Cow's milk 85% H_2O

Dry weight 20% Fat

20% Proteins

60% Sugars - largely lactose

also have sweat, sebaceous, scent, and musk glands

Hair: bodies of mammals typically covered with hair, which has no structural homology in other vertebrates.

- perhaps developed before a scaly covering lost in Therapsid reptiles
- consist of dead epidermal cells that are strengthened by keratin

Fat and energy storage:

Fat and adipose tissue but are of vital importance as:

- energy storage
- a source of heat and water
- thermal insulation.

Lives of many mammals punctuated by times of crisis when food is in short supply or energy demands are usually high

Circulatory System

Highly efficient system with four-chambered heart acting as a double pump.

The RIGHT side receives venous blood from the body and pumps it to the lungs for oxygenation.

The LEFT side receives oxygenated blood from the lungs and pumps it to the body.

Erythrocytes biconcave, enucleated disk as possible mechanisms for increased oxygen-carrying capacity.

Respiratory System

Lungs are large and, together with the heart, virtually fill the entire thoracic cavity.

Movements of air into and out of the lungs and volume of exchange due primarily to muscular diaphragm.

Reproductive System

Both ovaries are functional and the ova is fertilized in the oviducts.

Embryo develops within the uterus and lies within a fluid-filled amniotic sac

Nourishment for embryo comes from the maternal blood stream via placenta

Male testes typically contained within the scrotum outside the body cavity

Brain

Enlargement of the brain's cerebral hemisphere.

Neopallium - functions as center for sensory stimulus and initiation of motor activity

Sense Organs

Sense of smell acute as a result of development of the turbinate bones

Olfactory lobes enlarged in carnivorous and insectivores but lost in porpoises and dolphins. Hearing highly developed due to three middle ear bones: malius, incus, stapes, and external pinnae.

Tapetum lucidum - reflective structure within choroid that improves night vision by reflecting light

Vibrissae - tactile hairs/whiskers in the muzzles and lower legs of some mammals.

Digestive System

Salivary glands are present -specialized in anteaters: mucilaginous material makes the tongue sticky

Musculature System:

Limb and trunk musculature highly plastic

Variations for high speed locomotion.

The Skeleton

Basic changes from Reptiles to Mammals

- simplification of skeletal elements
- reduction in the size and number of bones
- limbs and girdle systems simplified
- axial skeleton becomes more rigid
- ossification of large parts of the skeleton
- development of epiphysis and diaphyses

Skull :

- increased brain case size
- sagittal and lambdoidal crest increased
- temporal muscle origin
- zygomatic arch protects eyes and provides an origin for the masseter
- turbinal bones within the nasal cavity (improved smell/saturation of air)
- foramina allow passage of cranial nerves
- 3 middle ear ossicles
- dentary bone articulates directly with the squamosal
- hyoid apparatus supports trachea, larynx, and base of the tongue

Teeth:

- Heterodonty - specialized for feeding/diet
- Originate in the premaxilla, maxilla, and dentary
- Dentine covered by enamel

Axial skeleton - limbs and girdles:

- five well-differentiated vertebrae: cervical, thoracic, lumbar, sacral, caudal

- sternum well developed to form a rigid rib cage
- limb motion generally restricted to fore-aft directions in distal joints; more solid hip and shoulder attachments
- pelvic girdle has characteristic shape: illium projecting forward and ischium and pubis back - all solidly fused
- standard pattern of bones in manus and pes (hand and foot) 2-3-3-3-3

Diagnostic Mammalian Traits:

Pelage = hair

Ears

Mammary glands

Diaphragm

Left aortic arch

Enucleated erythrocytes

3 middle ear bones

Single dentary

Dentary/squamosal jaw articulation

Mammal - a hairy, endothermic, homeotherm which, in most cases, bears live young which are nursed from mammary glands

Studying the evolution of Mammals:

- helps us understand where and why they are distributed
- understand evolution
- understand our past

Geologic Time

Era =» Period =» Epoch

Eras:

Precambrian or Protozoic- to 570mya

First life - algae, bacteria, worms

Paleozoic (old) - 570 - 225mya

Age of fishes

Mesozoic (middle) - 225 - 65mya

Age of reptiles ~20 reptilian orders

Cenozoic (recent) - 65mya to present

Age of mammals ~30 mammalian orders

Mesozoic mammals tended to be somewhat insignificant - limited fossil evidence indicated holding to conservative mouse-like form and quadrupedal locomotion

Dramatic adaptive burst following the extinction of the dinosaurs

Why so many mammals in such a short time?

Plate tectonic and continental drift
Mesozoic - Pangea 230 mya
Laurasia - Europe and Asia - northern
Gondwanaland - India, S. America, Africa, Antarctica, Australia, southern
When mammals arose, continents were fairly close together
- pieces breaking off with groups of mammals
- different conditions evolved different mammals
  - geographic isolation

Mammals evolved from Synapsid reptiles
1. old reptile group
2. past its peak prior to dinosaurs
3. competition from other reptiles
4. named for skull type
5. two important orders - Pelycosauria (primitive) and Therapsida (advanced)

Evolution of the Skull

Synapsid (one window) skulls:
1. allow jaw muscles to bulge
2. more surface area for muscle attachment
3. lighter skull
4. opening thickened around edges due to pressures

Order Pelycosauria

1. primitive order that gave rise to the Therapsids
2. first group to depart radically from the basic reptilian design = 300mya
3. "bowl lizards" pelvic structure allowed organs to be carried off of the ground- increased agility
4. many changes in size, teeth, skulls, jaw musculature
5. many forms evolved
6. may have had increased body temperature, appetites, and feeding effectiveness

Order Therapsida; Infraorder Cynodontia (dog tooth)

- basal stock for mammals
- about 2m in length
- highly cursorial (limbs rotated)
- well developed secondary palate
- reduction in number of lumbar ribs = probable development of diaphragm
- elevated metabolism . . . Why?

- diaphragm present - efficient respiration, more active
- locomoter efficiency
- histiological bone examination - no growth rings, Haversian canals - cell deposition, vessel patterns
- location of fossils - 60° latitude, no good hibernation sites for some, must produce heat to survive
- homeothermy - insulation, sweat glands = hair?? Evidence for hair vibrissae pits on cynodont skulls
- heterodonty - came with increased locomoter efficiency - probably with increased oxygen uptake

Increased locomoter efficiency and increased oxygen uptake allowed and aided in chasing and securing prey = differentiation in teeth

Greater mastication prior to swallowing - increased surface area for better enzyme action

Dentary-Squamosal jaw articulation developing with reduction of the quadrate and articular - become free to aid in transmission of sounds and vibrations.

Secondary palate increased surface area of the nasal cavity

- allowed for simultaneous mastication and respiration

- warm, moisturize, and clean air

- increase sense of smell - chemical smells/pheromones (vomeronasal organ)

Masseter muscle has essentially the same attachment as in modern mammals - insertion on lateral surface of dentary and originates on the zygomatic arch; enhanced control of transverse jaw movement.

First Mammals - after Cynodonts

~ 10cm long

20-30gm

long snouts

rows of complex teeth - probably insectivorous

partially arboreal

probably nocturnal - favored by endothermy

Early mammals were small - at least an order of magnitude smaller than Cynodonts

- arose in Triassic

- stayed small with Cretaceous (140mya)

Why stay small?

- competition from moderate-sized non-dinosaur reptiles (turtles, crocodiles) - probably not, lived in different habitats

- competition from small immature dinosaurs - carnivorous when young
- lacked sophisticated evaporative cooling mechanism

Mesozoic Mammalian Radiation

Current evidence indicates that mammals probably evolved monophyletically from cynodont reptiles

Early mammals displayed structural features that distinguish them from even the most advanced cynodonts:

1. In species of like body size, the morganucodontid brain was three or four times larger than that of even the most advanced therapsids, a reflection perhaps of greater neuromuscular coordination and improved auditory and olfactory acuity.
2. The condyle of the dentary bone fit into the glenoid fossa of the squamosal bone.
3. The cheek teeth were differentiated into premolars and molars, and the premolars were probably preceded by deciduous teeth.
4. Chewing was on one side of the jaw at a time, and the lower jaw on the side involved in chewing followed a triangular orbit as viewed from the front
5. During chewing, the inner surface of the upper molars sheared against the outer surface of the lower molars.
6. The cochlear region of the skull was far larger and more conspicuous ventrally than in cynodonts.
7. Body weight, probably 20 to 30 grams, was an order of magnitude smaller than in any Middle Triassic cynodont.
8. The pelvis was esscntially mammalian, with a rodlike ilium and a small pubis .
9. As part of a series of specializations allowing rotary head movement, the dens of the axis was large and protuberant and fit into the atlas.
10. The thoracic and lumbar vertebrae arched dorsally, the thoracic vertebrae had narrow, posteriorly directed neural spines, and the lumbar vertebrae bore dorsally directed neural spines.

Early radiation best described as a dichotomy between two early groups, the Kuehneotheriidae and Morganucodontidae.

Morganucodontidae - basic triconodont molar - may have evolved triconodons, docodonts, and monotremes. Kuehneotheriidae- triangular molars - may have given rise to the therians (symmetrodonts, pantotheres, marsupials, and eutherians)

Increasing evidence indicates more complex relationships among early mammals - only the monotremes, marsupials, and eutherians survive today.

Order Triconodonta - late Triassic early cretaceous- one of the oldest primitive prototherians - predaceous - largest = house cat size - heterodont - 14 teeth in dentary - canines large - molar cusps 3, arrangement in front-to-back row

Order Docodonta - late Jurassic- roughly quadrate teeth -cusp not aligned anterioposteriorly

Order Symmetrodonta - late Triassic to late Cretaceous- probably predaceous - 3 fairly symmetrical cusp

Order Multituberculata - first appeared in the late Jurassic period to Tertiary- first mammalian herbivores - wide spread in both the old and new worlds - Ecologically equivalent to rodents - strongly built lower jaw with attachment for powerful jaw muscles - 2 or 3 incisors - diastema in front of premolars, 3 parallel cuspules - olfactory lobes enlarged

It has been generally accepted that eutherians and metatherian mammals evolved from the Order Pantotheria

- profile of the ventral border of the dentary bone is interrupted by an angular process
- the lower molar has a posterior "heel" which is separated by the talonid
- the trigonid section of the pantotheric lower molar and the triangular upper molar resemble the corresponding teeth of some primitive eutherians and metatherians

During the Cretaceous, land dwellers were banned from intercontinental movement by oceans and seaways

- populations of mammals on different continents evolved in isolation under different environmental conditions
- earliest known marsupials are from late Cretaceous Canada, Westerns N.A. and Peru.
- mammalian radiation coincided with a burst of flowering plants (Angiosperms), Lepidoptera (moths and butterflies), Isoptera (termites), and Coleoptera (beetles).

Order Insectivora

Refers to the diet of many, but . . . taxonomic uncertainty and disagreements on classification are causing many problems - the order serves as a convenient "catch-all"

Most primitive eutherian order

Third largest order with ~77genera and 400spp.

Rodents ~1700spp.

Chiroptera ~850spp.

Distributed through most of both hemispheres except Australian region, northern part of South America, and polar regions

Originally thought to have evolved in Old World (Europe and Asia) and moved into the New World

- earliest fossil Insectivores (Batodon) from mid-Cretaceous North America ~100mya

- oldest members of clearly recognized families - soricids and talpids - from Eocene ~50mya

"Grab-bag" of forms making it difficult to form subclassifications

- many generalized forms
- some could be lumped
- some may be considered as separate orders
- Butler (1972) ". . . any fossil eutherian not closely related to one of the other orders is classified in the order Insectivora."

Order Macroscelidea - Elephant shrews

Order Scandentia - Tree shrews

General characteristics:

- usually small
- long narrow snout
- 5 clawed digits
- usually short, close-set fur
- anterior vena cavae paired
- pinnae small to absent
- minute eyes - some covered with skin
- scrotum when present anterior to penis
- most insectivorous
- terrestrial, fossorial, semiaquatic
- plantigrade - heels touch ground when walking

General cranial traits:

- small brain case with smooth cerebral hemisphere
- no auditory bullae - ring-shaped tympanic bone
- jugal reduced or absent
- zygomatic arch absent in some
- usually enlarged and specialized incisors with sharp shearing cusps
- canines usually reduced
- some genera retain tribosphenic molars

Order Chiroptera

Second largest order of mammals with ~170 genera and 850 species

Characterized as the only mammal to have evolved true flight

Represents the most poorly understood/misunderstood groups of mammals

Relatively recent biological research has revealed:

- extraordinarily complex social behavior, including harems maintained by males and complex vocalizations
- coordinated neuromuscular and behavioral adaptations allowing detailed perception of prey and their environments by the use of sound
- unsurpassed ability to conserve daily energy or survive through periods of stress by drastic reduction of metabolic rates

Bats have nearly cosmopolitan distribution, being absent only from the arctic and polar regions and from isolated oceanic islands

Frequently abundant members of temperate faunas but reach their highest densities and greatest diversities in tropical and subtropical areas.

Bats occupy a number of terrestrial environments, including:

- temperate, boreal, and tropical forest
- grasslands
- chaparral
- deserts
- also man-made structures afford excellent roost sites and agricultural areas having high insect abundance.

Paleontology

Because of their small size, ability to fly, delicate structure, and greatest abundance in tropical areas where fossilization rarely occurs, but fossils are rare

· Icaronycteris index - from Eocene beds in Wyoming the oldest known undoubtable bat material

· first described by Jepsen in 1966· claws on the first 2 digits of the hand· fairly short, broad wings

· Other fossils prove insectivory during Eocene - moth scales in gut

Fossil record shows little change in some families since Eocene, indicating good adaptations to their particular environments - a sharp contrast to Oligocene horses which were sheep-sized and 3-toed

Oligocene Tadarida were nearly identical to present-day members of the Molossidae family

Paleocene origins of bats seems probable, followed by a late Cretaceous divergence from insectivorous stock

Order Primates

Primates represent the 7th largest order with 51 genera and 168sp

- 16 genera and 50 species in the New World
- often considered to be the most important mammals (ego)
- no one denies that modern man is primate yet . . .
- few people understand why man is classified with animals such as the tree shrew, loris, and aye aye
- anthropologist have difficulty in defining what man is
- primates (living or fossil) are defined by an overall pattern
- The fossil record is from the:
- Palocene and Oligocene of North America and Europe
- Miocene of Africa and Europe
- Pleistocene and recent of South America, Asia, Africa, and Madagascar

Order Scandentia - tree shrews as primates

5 genera, ~16sp

- Appearance - don't look like primates but internally primate-like, can't base on superficial appearances
- Multiple breast pairs - have them but reduced in number, 1-3 pairs primitive condition
- Muzzle too long - but shorter than most insectivores - baboon's long
- External and middle ear - external human-like, middle ear primate-like
- Olfaction - rely on hearing, primates are more vision oriented with increase in brain section for vision
- Vision - poor binocular but retina is primate-like
- Throat-chest scent glands - lower primates scent mark
- Nonprehensile feet and hands / not opposable but is moveable
- No baculum or os clitoris bones - neither do humans
- Multiple births - primates don't usually, but more primitive groups do
- Maturation period too short - mature quickly rather than slowly - lots of differences in recognized primates
- Brain size relatively large in proportion to the body
- Blood/serological test closer to primates
- Social behavior poorly developed

Primate Pattern

Centers around adaptation to tree life and arboreal existence, secondarily evolved for life on the ground

Qualities to exploit forest canopies:

- behavioral plasticity - adaptation allows change at a second's notice
    - general anatomy - "plain"
    - neurologically complex
    - individuality - nervous system allows quick thinking
    - large cerebral centers devoted to hands, thumbs, and vocal areas
- grasping extremities - holding on in trees
    - opposable thumbs and toes
    - pentadactyl
    - grip powerful and precise, claws reduced to nails and more sensitive, friction ridges/callouses
- good vision - eyes in the front of the head
    - nose reduced or rostrum dropped to move nose out of the field of view
- mobile limbs - propelled through the trees ability to hold body erect - stand up to free the hands

Xenarthra (Edentata), Philodota, and Tubulidentata all share a major structural trend, the loss or simplification of dentition

Xenarthrans (armadillos, sloths, anteaters): underwent tertiary radiation in South America

Pholidota (pangolins) and Tubulidentata (aardvark): Old World groups, each of which has conservatively maintained a single structural plan.

All share a series of distinctive morphological features:

- extra zygopophysis-like articulation (Xenarthrous) which brace the lumbar vertebrae
- incisors and canines absent
- cheek teeth, when present, lack enamel and each has a single root - continuous growing
- brain case usually long and cylindrical
- the hind foot is usually five-toed
- forefoot has two or three prominant toes with large claws

Major Xenarthrous structural trends are toward a reduction and simplification of the dentition, specialization in the limbs for climbing or digging, and rigidity of the axial skeleton.

Order Carnivora - Predators

An ancient, profitable (and honorable) occupation

Appeared in the early Paleocene (Creodonts) before most of the recent mammalian orders

Probably evolved in response to the food resources offered by expanding array of herbivores

Most recent carnivores are predaceous and have a remarkable sense of smell

· many carnivorous

· may be omnivorous - bears

· may be specialized - cats

Cursorial abilities may be limited - Ursidae, Procyonidae

well developed - cheetah, canids

General Characters:

- larger brain case
- good sense of smell
- well-developed canines
- small incisors - usually 3/3
- shearing and crushing teeth
- many have carnassial pair P^4/m_1
- cheek teeth vary
- "long faced" - P 4/4, m 2/3 - dogs, bears
- "short faced" - P 2/2, m 1/1 - cats
- transverse mandibular fossa - good for biting, not for grinding
- reduced or absent clavicle
- plantigrade or digitigrade
- well-developed scent glands

Order Cetacea (Greek = whale)

Mysticeti - baleen whales 11spp.

Odontoceti - toothed whales 67spp.

Noted as being mammals that are most fully adapted to aquatic life

Fossil information indicates cetaceans are an old and successful group:

- swimming ability
- capacity to echolocate in many
- social behavior

Order Rodentia (Greek "to gnaw")

About half of all mammals currently alive are rodents

Plants are the most abundant food source - rodents adapted as herbivores to take advantage of this food supply

30 families

418 genera

~1750 species

Why so successful?

- nearly cosmopolitan in distribution

- exploit a broad spectrum of foods
- important members of most terrestrial faunas
- often reach extremely high population densities
- small size to utilize shelters and escape predation
- high fecundity - large number of offspring, some survive
- rapid population turnover - natural selection operates quickly

Taxonomic relationships are difficult to understand

Convergence - distantly related but look alike

Heteromyidae and Dipodidae - morphologically similar in utilizing dry habitats

Divergence - closely related but look nothing alike

Geomyidae related to Heteromyidae - gophers fossorially adapted/ kangaroo rats long legged and saltatorial

Parallelism - closely related and pursue similar modes of life

Muridae and Cricetidae - Old and New World rodents frequently lumped together into the same family - look very much alike and are clearly related through the fossil record

Evolutionary radiation has allowed Rodentia to occupy niches filled by other orders

South America species resemble:

rabbits = *Hydrochoerus* sp. (Capybara)

antelope = *Cavia* sp. (Cavy)

Size extremely variable:

Largest: South America *Hydrochoerus* sp. - Capybara ~100 lbs

North America *Castor canadensis* - beaver ~50 lbs

Smallest: *Baiomys* sp.

Perognatus flavus

Reithrodontomys humilis ~5g

Fossil record for rodents is not very good due to small and fragile bones

- Oldest from Paleocene of North America
- Ischyromyidae - represented by only a few teeth
- *Eumegamys* - largest known rodent about size of hippo, skull ~2 ft long
- *Castoroides* - Pleistocene beaver from Mississippi river valley ~7 ft long

Early divergence of suborders based on mandibles - Split by Tullburg (1899) and Wood (1985)

Sciurognathi - the angular process of the dentary bone originates in the plane that passes through the alveolus of the incisor and is ventral to the alveolus

Hystricognathi - the angular process originates lateral to the vertical plane of the alveolus. Most distinguishing trait for the order are based on dental characters:

Incisors

- single pair in each jaw
- roots and pulp cavities open and evergrowing
- outside enamel, inside dentine rub against each other - differential wear produces chisel edge
- good for gnawing, grasping and holding, piercing

No Canines - long diastema

Cheek Teeth

- highly variable
- premolars may or may not be present
- some ever-growing, some rooted

Dental formula reduced to a maximum of 22 teeth

1/1 0/0 2/1 3/3 = 22

Dentition best suited for herbivory but varies

Onychomys - carnivorous

Dipodomys - omnivorous

Rodents have been broken out into four taxonomic groups based on jaw musculature and mandible/skull articulation

Protrogomorphs - represents the primitive condition

Sciuromorphs - squirrel-like

Hystricomorphs - porcupine-like

Myomorphs - mouse-like

Protrogomorphs

- Masseter muscles originate entirely on the zygomatic arch toward the placement of the origin of at least one division of the masseter on the rostrum
- represented only by *Aplodontia rufa* (Aplodontidae)

Sciuromorphs

- The insertion of the anterior part of the lateral masseter is shifted onto the anterior surface of the zygomatic arch and the adjacent part of the rostrum - improved gnawing and grinding
- The temporalis muscle is relatively large and the coronoid process is moderately well developed
- occurs in Sciuridae and Castoridae

Hystricomorphs· The origin of the medial masseter from the zygomatic arch to an extensive area on the side of the rostrum· passes through the often greatly enlarged infraorbital foramen· occurs in Dipodidae and most Hystricognaths

Myomorphs

- The anterior part of the lateral masseter originates on the highly modified anterior extension of the zygomatic arch and the anterior part of the medialmasseter originates on the rostrum and passes through the somewhat enlarged infraorbital foramen
- occurs in many members of the Sciurognathi, including all members of the family Muridae

Complicated jaw action allows the lower cheek teeth to move transversely or anterioposteriorly against the upper teeth, producing a crushing and grinding action.

Cheek teeth and incisors perform distinctively different functions - musculature is required to move the lower jaws into power positions for teeth to function - "division of labor"

Order Lagomorpha

Rabbits, hares, and pikas are not a very diverse group but are important members of many terrestrial communities, and are nearly cosmopolitan in distribution and were only absent from Antarctica, Australia, and southern South America

Two families with 13 genera and 80 species

- Leporidae - rabbits and hares
- Ochotonidae - pikas

Taxonomic origins are from the Paleocene of China - appear to share a common origin with rodents within the Paleocene order Anagalida

- Leporidae underwent most of it's early evolution in the Oligocene and Miocene of North America
- Ochotonidae originated in Eurasia and developed largest number in North America

Conservatism in evolutionary design may be related to the limitations of their functional position as "miniature ungulates" - direct competition with Artiodactyla may have limited lagomorphs to a single limited adaptive zone

Ungulates

Not a taxonomic term but is a general term for a large group of cursorial forms

The name implies that a hoof is present

Two Orders:

Perissodactyla - horses, rhinos, tapirs - "odd toed"

Artiodactyla - pigs, peccaries, hippos, camels, deer, antelope, cattle, sheep, goats - "even toed"

Came from the Condylarthra - a diverse group of ungulate-like mammals that existed from Cretaceous to the Oligocene

Exceptional cursorial ability - Why?

- evolved at same time Miocene grasslands were developing
- open country
- needed speed to avoid predators in the absence of trees

Perissodactyls

- reached peak in Eocene with their greatest diversity
- declined through Miocene
- few alive today

Artiodactyla

- peaked in Miocene following adaptive radiation
- may have out-competed Perissodactyls

Cursorial Specializations

Two factors determine speed:

- length of stride
- rate of stride (#/time)

Length of stride:

1) Lengthen the limb

- metacarpals and metatarsals elongate and fused - cannon bone
- other limb bones elongated

2) Loss or reduction of clavicle

- loss of clavical frees scapula and shoulder joint from a strongly fixed attachment to axial skeleton
  - scapula free to pivot and rotate about a point near its center allowing leg to move farther during stride
  - absorbs shock of foot striking ground and not transfered to entire body

3) Flexion of the spine

- well developed muscles attached to spine
- flexed when legs under body , extended when legs outstretched

Increased rate of stride:

1) Extra joints

- total speed of foot depends in part upon the speed of each joint and the number of joints
  - each new joint = increase foot speed
  - greater number of joints moving in the same plane also increases the speed of the limb

How done?

- Digitigrade/Unguligrade - lift head off ground
- Phalanges and metapodials
- Scapula free
- Flexion of spine

2) Specialization in musculature

- proximal migration of muscle masses
- gets weight away from extremites
- concentrates center of gravity
- insertion points migrate proximally - the closer the muscle to the insertion point the faster the response
- distal parts lighter (fewer muscles) - less energy needed to start/stop - oftern associated with a reduction in the bones
- plane of movement limited to one - fewer muscles needed

3) Ungulate "ankle/wrist" specializations

Feeding Specializations for Herivorous Diets:

1) Cheek teeth

- large teeth with complex occlusal surfaces
- premolars molariform - hypsodont
- diastema
- elongated jaw for broader teeth

2) Digestion of plant material

- lack enzymes to digest cellulose - uses bacteria
- retention time varies but is long
- 30 - 45 hours in Perissodactyles
- 70 - 100 hours Artiodactyls

Artiodactyls more efficient digesters - may live longer in areas with short food supply

Microflora and Fauna:

- in caecum of Artiodactyls
- # bacteria est. 10^{10} - divide every 20 min.
- protozoa est. 10^{6} - divide every 2-3 days
- originally thought to only digest cellulose
- Also
- synthesizes vitamin B
- makes amino acids, simple proteins
- split urea into ammonia, fix N to make amino acids
- decompose complex carbohydrates, convert to fatty acids
- produce large quantities of CO_2 and Methane

Ruminants = Cud chewers

General pattern of stomach activity:

- fill stomach rapidly and retire to safe spot to digest
- stomach large with 4 chambers
- rumen - paunch
- reticulum - honeycomb tripe
- omasum - psalterium
- abomasum - stomach of other mammals
- food swallowed into rumen and churned with micro flora and fauna
- regurgitated to mouth (cud) and chewed to mix with micro flora/fauna
- reswallowed
- mixed between rumen and reticulum
- enters omasum
- compressed and mixed more, 60 - 70% H_2O absorbed

CLASS OF VERTEBRATE MAMMALS

Mammals (formally Mammalia) are a class of vertebrate, air-breathing animals whose females are characterized by the possession of mammary glands while both males and females are characterized by sweat glands, hair and/or fur, three middle ear bones used in hearing, and a neocortex region in the brain. Mammals are divided into three main infraclass taxa depending how they are born. These taxa are: monotremes, marsupials and placentals. Except for the five species of monotremes (which lay eggs), all mammal species give birth to live young. Most mammals also possess specialized teeth, and the largest group of mammals, the placentals, use a placenta during gestation. The mammalian brain regulates endothermic and circulatory systems, including a four-chambered heart.

There are approximately 5,400 species of mammals, distributed in about 1,200 genera, 153 families, and 29 orders (though this varies by classification scheme). Mammals range in size from the 30–40 millimeter (1- to 1.5-inch) Bumblebee Bat to the 33-meter (108-foot) Blue Whale. Mammals are divided into two subclasses: the Prototheria, which includes the oviparous monotremes, and the Theria, which includes the placentals and live-bearing marsupials. Most mammals, including the six largest orders, belong to the placental group. The three largest orders, in descending order, are Rodentia (mice, rats, porcupines, beavers, capybaras, and other gnawing mammals), Chiroptera (bats), and Soricomorpha (shrews, moles and solenodons). The next three largest orders include the Carnivora (dogs, cats, weasels, bears, seals, and their relatives), the Cetartiodactyla (including the even-toed hoofed mammals and the whales) and the Primates to which the human species belongs. The relative size of these latter three orders differs according to the classification scheme and definitions used by various authors.

Phylogenetically, Mammalia is defined as all descendants of the most recent common ancestor of monotremes (e.g., echidnas and platypuses) and therian mammals (marsupials and placentals). This means that some extinct groups of "mammals" are not members of the crown group Mammalia, even though most of them have all the characteristics that traditionally would have classified them as mammals.

These "mammals" are now usually placed in the unranked clade Mammaliaformes. The mammalian line of descent diverged from an amniote line at the end of the Carboniferous period. One line of amniotes would lead to reptiles, while the other would lead to synapsids.

According to cladistics, mammals are a sub-group of synapsids. Although they were preceded by many diverse groups of non-mammalian synapsids (sometimes misleadingly referred to as mammal-like reptiles), the first true mammals appeared in the Triassic period. Modern mammalian orders appeared in the Palaeocene and Eocene epochs of the Palaeogene period.

MAMMALIAFORMES

Mammaliaformes ("mammal-shaped") is a clade that contains the mammals and their closest extinct relatives. Phylogenetically it is defined as a clade including the most recent common ancestor of *Sinoconodon*, morganuconodonts, docodonts, Monotremata, Marsupialia, Placentalia, extinct members of this clade, and all of its descendants. The precise phylogeny is disputed due to the scantness of evidence in the fossil record. However, it is thought that the Mammaliaformes were of three major groups: Allotheria, the longest extinct lineage of pre-mammals; Docodonta, including close relatives such as Morganucodonta; and Symmetrodonta, the most basal of modern mammals. Mammaliaformes radiated from Cynodontia. The Probainognathians of the Eucynodont clade probably evolved into the early mammaliaformes, but the branch Allotheria was so different that they may have come from an entirely different group of cynodonts.

Early mammaliforms were generally shrew-like in appearance and size, and most of their distinguishing characteristics were internal. In particular, the structure of the mammaliform (and mammal) jaw and arrangement of teeth is nearly unique. Instead of having many teeth that are frequently replaced, mammals have one set of baby teeth and later one set of adult teeth which fit together precisely. This is thought to aid in the grinding of food to make it quicker to digest. Being warm-blooded requires more calories than "cold-blooded" animals, so quickening the pace of digestion is a necessity. The drawback to the fixed dentition is that worn teeth cannot be replaced, as was possible for the reptilian ancestors of mammaliforms. However, small mammals generally being very short-lived compared to reptiles of the same size, this was not much of a problem during the early phase of their evolution, in which the trait was set. Early mammaliaformes were probably nocturnal. Mammaliforms have several common structures. Most importantly, mammaliforms have highly specialized molars, with cusps and flat regions for grinding food. This system is also unique to mammals, although it seems to have evolved convergently in pre-mammals multiple times.

Lactation and fur, along with other characteristically mammalian features, are also thought to characterize the Mammaliaformes, but these traits are difficult to study in the fossil record. The fossilized remains of *Castorocauda lutrasimilis* are a unique exception. Some non-mammal mammaliformes still retain reptile-like traits. Some mammaliformes had reptile-like locomotion. Furthermore, these mammaliformes still had some bones on their lower jaw seen in reptiles.

Monotreme

Monotremes are mammals that lay eggs (Prototheria) instead of giving birth to live young like marsupials (Metatheria) and placental mammals (Eutheria).

They are conventionally treated as comprising a single order Monotremata, though a recent classification proposes to divide them into the orders Platypoda (the Platypus along with its fossil relatives) and Tachyglossa (the echidnas, or spiny anteaters). The entire grouping is also traditionally placed into a subclass Prototheria, which was extended to include several fossil orders but these are no longer seen as constituting a natural group allied to monotreme ancestry. A controversial hypothesis now relates the monotremes to a different assemblage of fossil mammals in a clade termed Australosphenida.

General Characteristics

Like other mammals, monotremes are warm-blooded with a high metabolic rate (though not as high as other mammals, see below); have hair on their bodies; produce milk through mammary glands to feed their young; have a single bone in their lower jaw; and have three middle-ear bones. Monotremes were very poorly understood for many years, and to this day some of the 19th century myths that grew up around them endure. It is still sometimes thought, for example, that the monotremes are "inferior" or quasi-reptilian, and that they are a distant ancestor of the "superior" placental mammals. It now seems clear that modern monotremes are the survivors of an early branching of the mammal tree; a later branching is thought to have led to the marsupial and placental groups. In common with reptiles and marsupials, monotremes lack the connective structure (corpus callosum) which allows communication between the right and left brain hemispheres in placentals.

The key anatomical difference between monotremes and other mammals is the one that gave them their name; *monotreme* means 'single opening' in Greek and comes from the fact that their urinary, defecatory, and reproductive systems all open into a single duct, the cloaca. This structure is very similar to the one found in reptiles. Monotremes and marsupials have a single cloaca (though marsupials also have a separate genital tract), while placental mammal females have separate openings for reproduction, urination, and defecation: the vagina, the urethra, and the anus.

Monotremes lay eggs. However, the egg is retained for some time within the mother, who actively provides the egg with nutrients. Monotremes also lactate, but have no defined nipples, excreting the milk from their mammary glands via openings in their skin. All species are long-lived, with low rates of reproduction and relatively prolonged parental care of infants. Infant echidnas are sometimes known as *puggles,* referencing their similarity in appearance to the Australian children's toy designed by Tony Barber. The same term, though not generally accepted, is popularly applied to young platypuses as well.

Extant monotremes lack teeth as adults. Fossil forms and modern platypus young have "tribosphenic" molars (with the occlusal surface formed by three cusps arranged in a triangle), which are one of the hallmarks of extant mammals. Some recent work suggests that monotremes acquired this form of molar *independently* of placental mammals and marsupials, although this is not well established. The jaw of monotremes is constructed somewhat differently from that of other mammals, and the jaw opening muscle is different. As in all true mammals, the tiny bones that conduct sound to the inner ear are fully incorporated into the skull, rather than lying in the jaw as in cynodonts and other pre-mammalian synapsids; this feature, too, is now claimed to have evolved independently in monotremes and therians, although, as with the analogous evolution of the tribosphenic molar, this is disputed.

The external opening of the ear still lies at the base of the jaw. The imminent sequencing of the Platypus genome should shed light on this and many other questions regarding the evolutionary history of the monotremes. The monotremes also have extra bones in the shoulder girdle, including an interclavicle and coracoid, which are not found in other mammals. Monotremes retain a reptile-like gait, with legs that are on the sides of rather than underneath the body. The monotreme leg bears a spur in the ankle region; the spur is non-functional in echidnas, but contains a powerful venom in the male Platypus.

Physiology

It is still sometimes said that monotremes have less developed internal temperature control mechanisms than other mammals, but recent research shows that monotremes maintain a constant body temperature in a wide variety of circumstances without difficulty (for example, the Platypus while living in an icy mountain stream).

Early researchers were misled by two factors: firstly, monotremes maintain a lower average temperature than most mammals (around 32 °C (90 °F), compared to about 35 °C (95 °F) for marsupials, and 37 °C (99 °F) for most placentals); secondly, the Short-beaked Echidna (which is much easier to study than the reclusive Platypus) maintains normal temperature only when it is active: during cold weather, it conserves energy by "switching off" its temperature regulation. Finally, poor thermal regulation has also been observed in the hyraxes, which are placental mammals.

Their metabolic rate is remarkably low by mammalian standards. The Platypus has an average body temperature of about 32 °C (90 °F) rather than the 37 °C (99 °F) typical of placental mammals. Research suggests this has been a gradual adaptation to harsh environmental conditions on the part of the small number of surviving monotreme species rather than a historical characteristic of monotremes. Contrary to previous research, the Echidna does enter REM sleep, albeit only when the ambient temperature of its environment is around 25 °C (77 °F). At the temperatures of 15 °C (59 °F) and 28 °C (82 °F), REMS is suppressed.

Taxonomy

The only surviving examples of monotremes are all indigenous to Australia and New Guinea, although there is evidence that they were once more widespread.

Fossil and genetic evidence shows that the monotreme line diverged from other mammalian lines about 150 million years ago and that both the short-beaked and long-beaked echidna species are derived from a platypus-like ancestor. Fossils of a jaw fragment 110 million years old were found at Lightning Ridge, New South Wales.

These fragments, from species *Steropodon galmani*, are the oldest known fossils of monotremes. Fossils from the genera *Kollikodon*, *Teinolophos*, and *Obdurodon* have also been discovered. In 1991, a fossil tooth of a 61-million-year-old platypus was found in southern Argentina (since named *Monotrematum*, though it is now considered to be an *Obdurodon* species). Molecular clock and fossil dating suggest echidnas split from platypuses 19–48 million years ago.

- Order Monotremata;
- Suborder Platypoda.
    * Family Ornithorhynchidae: platypus
    * Genus *Ornithorhynchus*
    * Platypus, *Ornithorhynchus anatinus*.
- Suborder Tachyglossa.
    * Family Tachyglossidae: echidnas
    * Genus *Tachyglossus*
    * Short-beaked Echidna, *Tachyglossus aculeatus*
    * *Tachyglossus aculeatus aculeatus*
    * *Tachyglossus aculeatus acanthion*
    * *Tachyglossus aculeatus lawesii*
    * *Tachyglossus aculeatus multiaculeatus*
    * *Tachyglossus aculeatus setosus*
    * Genus *Zaglossus*
    * Sir David's Long-beaked Echidna, *Zaglossus attenboroughi*
    * Eastern Long-beaked Echidna, *Zaglossus bartoni*
    * *Zaglossus bartoni bartoni*
    * *Zaglossus bartoni clunius*
    * *Zaglossus bartoni diamondi*
    * *Zaglossus bartoni smeenki*
    * Western Long-beaked Echidna, *Zaglossus bruijni*.

ADAPTATIONS OF MAMMAL CHARACTERISTICS

Mammals characteristics include numerous adaptations that enable them to survive in a wide range of environments. They live in nearly every habitat around the globe, from frigid polar regions, to turbulent seas, to dense tropical forests. Modern mammals range in stature from tiny field mice to massive whales and although various species may look drastically different, all mammals still share a unifying set of characteristics.

Some mammal characteristics—such as their hair, mammary glands, and three specialized middle-ear bones—are shared by no other groups of animals. Here we'll explore key facts and information about mammals so we can better understand important mammal characteristics.

Hair

Hair is one of the characteristics of a mammal that is unique to mammals—no other animals have true hair and all mammals have hair covering at least part of their body at some time during their life. An individual hair consists of a rod of cells that are reinforced by a protein known as keratin. Hair grows from skin cells called follicles. Hair can take on several different forms including thick fur, long whiskers, defensive quills or fearsome horns. Hair serves a variety of functions for mammals. It provides insulation, protects the skin, serves as camouflage and provides sensory feedback.

Some mammals have thick coats of fur that consist of two layers, a soft underfur and a coarse protective outer fur. Sea otters, for example, have this type of two-layered fur. In fact, sea otters have one the thicket coats of fur of all mammals, with more than 100,000 individual hairs per square centimeter. Since sea otters lack a layer of insulating blubber, they must compensate by having fur with superior insulation power. Cetaceans, in contrast, have a thick layer of insulating blubber and therefore have lost most of their hair over the course of their evolution. Some whales only have hair during their early development, while others retain sparse patches of hair on their chin or upper lip.

Mammary Glands

Mammals nurse their young with milk produced by mammary glands. Mammary glands, like hair, are a uniquely mammalian trait. Though present in both males and females, in most mammal species mammary glands only fully develop in females. The exception to this rule is the male Dayak fruit bat, which has mammary glands that produce milk to feed its young.

Mammary glands are modified and enlarged sweat glands that consist of ducts and glandular tissues that secrete milk through nipples. Young mammals obtain milk from their mother by feeding from her nipples. The milk provides the young with much needed protein, sugars, fat, vitamins and salts.

Not all mammals have nipples. Monotremes, which include echidnas and the platypus, diverged from other mammals early in their evolution. Monotremes lack nipples and instead secrete the milk produced by their mammary glands through ducts in their abdomen.

Lower Jaw Made of a Single Bone

Mammals differ from other vertebrates such as reptiles, birds and amphibians in that they have a single lower jaw bone that attaches directly to the skull. This bone is referred to as the dentary, due to the fact that it holds the teeth of the lower jaw. In other vertebrates, the dentary is one of several bones in the lower jaw and does not attach directly to the skull.

The structure of the lower jaw and the muscles that control it provides mammals with a powerful bite and enables them to use their teeth to cut and chew their prey. Mammal species have specialized teeth adapted to their particular diet. Cats, for example, have sharp teeth that enable them to tear meat while herbivores such as bison have broad teeth, well-suited for grinding plant material.

Diphyodonty

Diphyodonty is a pattern of tooth replacement in animals in which the teeth are replaced only once throughout the lifetime. Young mammals have a set of teeth that are smaller and weaker than their adult teeth. This first set of teeth, also known as the deciduous teeth, fall out and are later replaced by a set of larger, permanent teeth. Polyphydonty, in contrast to diphyodonty, is a term used to describe the pattern of tooth replacement in which teeth are continuously replaced throughout the lifetime of an animal. Toothed fishes, reptiles and many other non-mammalian vertebrates are polyphydonts.

Three Middle Ear Bones

Mammals have a unique arrangement of three bones in the middle ear. These bones—the incus, malleus and stapes, commonly referred to as the hammer, anvil and stirrup—are unique to mammals, no other animal group has them. The middle ear bones transmit sound vibrations from the tympanic membrane or eardrum to the inner ear and transforms them into neural impulses. The malleus and incus are derived bones that were once part of the lower jaw in mammal ancestors.

Endothermy

Mammals are endothermic which means they are capable of regulating their own body temperature so that it remains at a relatively constant temperature regardless of the temperature of the surrounding environment.

Diaphram

The diaphram is a layer of muscle located at the base of the ribcage that separates the thoracic cavity from the adominal cavity in mammals. Mammals are not the only vertebrates to posess a diaphram, amphibians and reptiles also have diaphragms or diaphragm-like structures. It should be noted that the anatomy of the diaphram and its position vary among the different classes of vertebrates.

Four-Chambered Heart

Like all vertebrates (and even all animals with a circulatory system), mammals have a muscular heart that contracts repeatedly to pump blood throughout the body's blood vessels. The heart serves to deliver oxygen and nutrients throughout the body and remove waste products. In general, the heart consists of multiple chambers (the number of chambers differs for the various animal groups). Two to four chambers may be present and there are two types of chambers, the atrium and ventricle (the atrium receives the blood returning to the heart while the ventricle pumps blood from the heart to the rest of the body).

The structural details of the heart differs among the various animal groups. Fish have the simplest heart structure of all vertebrates which consists of two chambers (one atrium and one ventricle). Amphibians and most reptiles have a three-chambered heart (two atria and one ventricle). Birds and mammals have a four-chambered heart (two atria and two ventricles).

The structure of a four-chambered heart offers greater efficiency than the three- and two- chambered heart structures. A four-chambered heart separates oxgenated blood coming from the lungs from the partially deoxygenated blood returning from the body to the lungs to be re-oxygenated. The prevention of mixing of these two streams of blood ensures that tissues receive oxygen-rich blood which in turn enables sustained muscle activity and helps in maintaining constant body temperatures.

DIVERSITY OF LAND MAMMALS

There is considerable change in the diversity of Texas mammals with geography. To illustrate this, species diversity has been depicted along a series of quadrats positioned along two transects that traverse the state (one stretching in a west to east direction from El Paso to Beaumont and another beginning at Dalhart in the northern part of the state and continuing southeastward to Brownsville). Species diversity exhibits a general decrease along the transect from El Paso to Beaumont. The lowest diversity is in the Blackland Prairies region (quadrats 12 and 13) and the highest is in the Guadalupe Mountains of the Trans-Pecos (quadrat 3). Major shifts in the diversity pattern are evident on either side of the Balcones Escarpment (between quadrats 10 and 12), and between the western portion of the Edwards Plateau (quadrat 4) and the Guadalupe Mountains in the Trans-Pecos (quadrat 3).

The pattern is much more irregular, without any general trend, along the north to south transect. Diversity is highest in the Escarpment Breaks of the High Plains (quadrat 5), the Balcones Canyonlands of the Edwards Plateau (quadrats 10 and 11), and the subtropical brushlands of the South Texas Plains (quadrat 17). Diversity along this transect is lowest in the Rolling Plains region (quadrat 7) and the coastal sands of the South Texas Plains (quadrat 15).

Species diversity can also be viewed in terms of habitat diversity and land area. To evaluate this, the diversity of Texas mammals was examined with respect to the 10 major vegetation regions in the state. The plot of the number of species in each vegetative region versus the log of the land area for that vegetative type. The regions of lowest mammalian diversity in Texas are in the eastern half of the state (Pineywoods, Gulf Prairies and Marshes, Post Oak Savannah, Blackland Prairies, and Cross Timbers region) and on the High Plains. Areas of highest mammalian diversity are in the Trans-Pecos, Edwards Plateau, South Texas Plains, and Rolling Plains.

Two important generalizations are evident about the diversity of Texas mammals. First, there is no strong correlation between land area of the vegetation regions and species diversity. For example, the High Plains region is slightly larger in area than the Trans-Pecos region yet it supports only about half as many species of mammals. Second, those natural regions of Texas where vegetative and topographic heterogeneity are the greatest provide a broader spectrum of potential mammalian habitats and thus support a greater number of mammalian species.

The Protochordata

Protochordate, any member of either of two invertebrate subphyla of the phylum Chordata: the Tunicata (sea squirts, salps, etc.) and the Cephalochordata (amphioxus). Like the remaining subphylum of the chordates, the Vertebrata, the protochordates have a hollow dorsal nerve cord, gill slits, and a stiff supporting rod, the notochord, the forerunner of the backbone. The protochordates differ chiefly from the vertebrates in not having a backbone. Recent protochordates are thought to have evolved from the same ancestral stock as that which gave rise to the vertebrates.

Two main theories have gained general acceptance as to how the vertebrates may have evolved. One theory proposes that the ancestral form was sessile (attached), perhaps like apterobranch but with an unspecialized larva. This larva adapted to an independent pelagic lifeand became sexually mature. Subsequently, the sessile stage was lost, and the vertebrates evolved from this free-swimming animal. The other, more recent theory postulates that the chordates evolved from a small fossil group called the mitrates.

Classification of Protochordata

Protochordata is comprised of three subphyla: Hemichordata, Urochordata and Cephalochordata. Hemichordate seems to be very remotely connected to urochordata, while urochordate and cephalochordate are found to be closely related to each other.

Ancestry of Chordates

The lower chordates and echinoderms show striking similarities in their development and it is suggested that the evolution of the ancestral chordates has taken place from the larvae of echinoderm like animals. The auricularia larva of echinoderms is a free swimming bilaterally symmetrical larva, and it very closely resembles the toronaria larva of hemichordates. It is also suggested that a modified auricularia larva probably became sexually mature and this neotenic larva could be the ancestor from which the chordates evolved. Another significant resemblance between echinoderms and hemichordates lies in the origin and arrangement of coelomic cavities. This phylogenetic relationship is unquestionable and it appears reasonable that echinoderms and hemichordates arose from a common ancestor. Thus the protochordates form a connecting link between the invertebrates and vertebrates.

PROTOCHORDATES, VERTEBRATE PHYLOGENY AND EMBRYOLOGY

Protochordates

Protochordates are an informal category of animals (i.e.: not a proper taxonomic group), named mainly for convenience to describe invertebrate animals that are closely related to vertebrates. This group is composed of the Phylum Hemichordataand the Subphyla Urochordata and Cephalochordata.

The Phylum Hemichordata consists of marine worms that share some, but not all of the characteristics of chordates. These animals have pharyngeal gill slits and a dorsal nerve cord, which is usually solid. The three body parts are proboscis, collar and trunk. What was once thought to be a notochord is no longer considered homologous. Acorn worms are examples of hemichordates.

The Urochordates and Cephalochordates are protochordates, but belong to the Phylum Chordata. Therefore, these animals will be discussed in the chordate section.

MOLECULAR CLASSIFICATION OF PLACENTALS

Molecular studies based on DNA analysis have suggested new relationships among mammal families over the last few years. Most of these findings have been independently validated by Retrotransposon presence/ absence data. The most recent classification systems based on molecular studies have proposed four groups or lineages of placental mammals. Molecular clocks suggest that these clades diverged from early common ancestors in the Cretaceous, but fossils have not been found to corroborate this hypothesis. These molecular findings are consistent with mammal zoogeography:

Following molecular DNA sequence analyses, the first divergence was that of the Afrotheria 110–100 million years ago. The Afrotheria proceeded to evolve and diversify in the isolation of the African-Arabian continent. The

Xenarthra, isolated in South America, diverged from the Boreoeutheria approximately 100–95 million years ago.

According to an alternative view, the Xenarthra has the Afrotheria as closest allies, forming the Atlantogenata as sistergroup to Boreoeutheria. The Boreoeutheria split into the Laurasiatheria and Euarchontoglires between 95 and 85 mea; both of these groups evolved on the northern continent of Laurasia. After tens of millions of years of relative isolation, Africa-Arabia collided with Eurasia, exchanging Afrotheria and Boreoeutheria.

The formation of the Isthmus of Panama linked South America and North America, which facilitated the exchange of mammal species in the Great American Interchange. The traditional view that no placental mammals reached Australasia until about 5 million years ago when bats and murine rodents arrived has been challenged by recent evidence and may need to be reassessed.

These molecular results are still controversial because they are not reflected by morphological data, and thus not accepted by many systematists. Further there is some indication from Retrotransposon presence/absence data that the traditional Epitheria hypothesis, suggesting Xenarthra as the first divergence, might be true. With the old order Insectivora shown to be polyphylectic and more properly subdivided (as Afrosoricida, Erinaceomorpha, and Soricomorpha), the following classification for placental mammals contains 21 orders:

- Clade Atlantogenata;
- Group I: Afrotheria.
    - * Clade Afroinsectiphilia
    - * Order Macroscelidea: elephant shrews (Africa)
    - * Order Afrosoricida: tenrecs and golden moles (Africa)
    - * Order Tubulidentata: aardvark (Africa south of the Sahara)
    - * Clade Paenungulata
    - * Order Hyracoidea: hyraxes or dassies (Africa, Arabia)
    - * Order Proboscidea: elephants (Africa, Southeast Asia)
    - * Order Sirenia: dugong and manatees (cosmopolitan tropical).
- Group II: Xenarthra.
    - * Order Pilosa: sloths and anteaters (Neotropical)
    - * Order Cingulata: armadillos (Americas)
- Clade Boreoeutheria;
- Group III: Euarchontoglires (Supraprimates).
    - * Superorder Euarchonta
    - * Order Scandentia: treeshrews (Southeast Asia).
    - * Order Dermoptera: flying lemurs or colugos (Southeast Asia)

  * Order Primates: lemurs, bushbabies, monkeys, apes (cosmopolitan), humans
  * Superorder Glires
  * Order Lagomorpha: pikas, rabbits, hares (Eurasia, Africa, Americas)
  * Order Rodentia: rodents (cosmopolitan).
- Group IV: Laurasiatheria.
  * Order Erinaceomorpha: hedgehogs
  * Order Soricomorpha: moles, shrews, solenodons
  * Clade Ferungulata
  * Clade Cetartiodactyla
  * Order Cetacea: whales, dolphins and porpoises
  * Order Artiodactyla: even-toed ungulates, including pigs, hippopotamus, camels, giraffe, deer, antelope, cattle, sheep, goats
  * Clade Pegasoferae
  * Order Chiroptera: bats (cosmopolitan)
  * Clade Zooamata
  * Order Perissodactyla: odd-toed ungulates, including horses, donkeys, zebras, tapirs, and rhinoceroses
  * Clade Ferae
  * Order Pholidota: pangolins or scaly anteaters (Africa, South Asia)
  * Order Carnivora: carnivores (cosmopolitan).

7

Development of Fishes

At the onset of the nineteenth century, the study of chondrichthyan development was already farther advanced than that of teleosts. However, the very features that had initially favoured the study of chondrichthyan fishes, e.g., large eggs and embryos, viviparity, and internal fertilization, subsequently worked against their further study. It was difficult to obtain access to early stages of development in viviparous species and there was no really effective way to maintain egg-laying species in captivity so as to obtain newly laid eggs.

The rate of development was slow. The large, meroblastic eggs were opaque, so that only external features of development and those details revealed by microdissections of embryos could be observed. In contrast, the introduction of artificial fertilization made the early stages of teleostean development readily available and the transparency of most teleostean eggs and embryos was well suited for the microscopic examination of external and internal features of development.

Accordingly, nineteenth century studies of chondrichthyan development can be divided into two phases. Phase One spanned the period 1800-1860. During this phase, research progressed along previously established lines with an emphasis on reproductive biology, the study of viviparous species and adaptations for viviparity, and the later stages of embryonic development. Phase Two spanned the period 1860-1911. Research in this phase was shaped by the introduction of new embryological and histological microtechniques to the study of chondrichthyans.

Research of the eighteenth century passed imperceptibly into the nineteenth century with little or no change of agenda. In addition to his magnum opus, Bloch published a set of separate papers in obscure journals, some of which dealt with chondrichthyan reproduction and development. Tilesius von Tilenau (1802) published a monograph on the egg cases of sharks and skates. It is noteworthy because it represents the first attempts to determine the chemical composition of egg cases, now known to be a type of "tanned" protein. In addition, he also reported on the urogenital anatomy and the reproduction and development of sharks and skates. The major significance of Mitchell's (1803) report on the anatomy of early (7-8 cm) pre-implantation shark embryos is that it is one of the earliest embryological papers and probably the first paper on fish embryology published in the United States.

Home's (1810) paper had two major aspects: one embryological and the other reproductive. He provided anatomical descriptions of the male and female reproductive system and described the structural changes that occur as mature individuals pass from a reproductively inactive to an active state. His study of the male urogenital system and claspers (modified caudal portion of each pelvic fin) represents the first step toward understanding the process of chondrichthyan insemination and fertilization. He gives a fine illustration of the claspers with their terminal processes expanded as they would be in copulation. He also describes the urogenital papilla, the terminal end of the male reproductive duct, but mistakenly refers to it as a "penis" and states that it is the intromittent organ.

Claspers were first reported by Aristotle, and their gross anatomy described by Lorenzini (1678) and Bloch (1782-1795). A controversy developed over the centuries concerning their function either as an intromittent organ or an organ for holding the female fast during copulation Further contributions were made by Davy (1839). He described the internal anatomy of the clasper that makes possible its function in sperm transfer. On the basis of his own work and a review of the literature, he concluded that the clasper functions both as a holdfast and intromittent organ. He also described the accessory clasper gland of skates and rays, a structure now known to produce prostaglandins (Lacy, personal communication).

Davy (1839) appears to be the first to have observed chondrichthyan sperm. In the electric ray (Torpedo) and a skate, he described sperm with an elongated head and flagellum. In vitro microscopic examination revealed that their motion was "serpentine and vibratory." Details on the process of spermatogenesis and the structure of the mature sperm came later. In 1871, Agassiz stated that when the claspers (of a skate) are rotated forward and upwards, an opening in them was brought opposite the spermatic duct (=urogenital papilla). The clasper could then be introduced into the oviduct and spermatic fluid could pass up the clasper and into the oviduct. Bolau (1881) appears to be the first to actually observe and describe the mating activity of sharks. The current view of insemination is that of Gilbert and Heath: "In mating, one clasper is flexed medially, inserted, and is anchored in the oviduct by a complex of cartilages at the clasper tip. Sperm pass from the urogenital papilla into the clasper groove and are washed into the oviduct by seawater and secretions expressed from a siphon sac." The actual cytological events of fertilization in chondrichthyans did not become known until Ruckert (1892b) demonstrated physiological polyspermy.

Embryology was the second aspect of Home's (1810) paper. He compared stages in development of the oviparous catshark, Scyliorhinus with the non-placental viviparous spiny dogfish shark, Squalus. He illustrated an advanced yolk sac embryo of the catshark in its egg case. He illustrated and described the slits that open in the egg case to permit the flow of seawater through the egg case during mid-late phases of development. He gave an excellent rendering of an oviduct of gravid Squalus with its ostium tubae, nidamental (shell) gland, and posterior uterus lined with longitudinal vascular ridges. Within the uterus, the "candle" is illustrated for the first time. It is an attenuated egg case that contains three developing eggs and a quantity of transparent jelly. Neonates of both the catshark and spiny dogfish are illustrated. To determine the physical and chemical properties of the oviducal jelly of the spiny dogfish, Home (1810) enlisted the aid of the chemist W. Brande. The jelly was extremely hydrophilic, being able to absorb many times its volume of water. Chemically the jelly was found to be neither gelatin nor albumen. It is interesting to note that Home, under the influence of Humphrey Davy, speculated on the requirements of developing fish eggs for aeration and how this process might be accomplished with different patterns of embryonic circulation.

Rathke was von Baer's successor at Konigsberg. His study of the "sexual apparatus" in fishes and other vertebrates was the first major step in understanding the development of the urogenital system, the differentiation of the gonads and accessory ducts, and the clarification of relationships in the vertebrate urogenital system. His monograph on the development of sharks and rays was primarily concerned with organogenesis and the microanatomy of the developing embryo. His comparative study of the development of the gill slits and arches forms the basis for homologizing the pharyngula phase of vertebrate development and was used extensively by the proponents of the biogenetic law later in the century.

John Davy, the brother of Humphrey Davy, published an account of the embryonic development of the electric ray Torpedo and a series of experiments on its electric organs. In his description of the female reproductive system, he noted that the absence of a nidamental (shell) gland accounts for the absence of an egg envelope in this fish. He presents an illustrated series of developmental stages which includes information on the electric organ. His earliest embryo was 18 mm long. It had short, external gill filaments but lacked an electric organ. The electric organ had begun its development in 28 mm embryos. Pectoral fin development began in 39-35 mm embryos, and by 62 mm the embryo looked like a miniature adult. Living full-term embryos were surgically removed from the female and maintained in sea water for up to six months. These neonates had functional electric organs. Activity was measured with a galvanometer. At six months, the young rays still had vestiges of an internal yolk sac and showed no interest in food. Davy suggests that they depend on yolk reserves during this period. Davy's (1834) contributions to the study of embryonic nutrition will be discussed in another section. Almost simultaneously, Leuckart (1836), a protege of Rathke, published a comparative study of the external gill filaments of several species of sharks and the electric ray. This paper has added interest because Leuckart's incorporation of an extensive series of measurements of embryonic structures represents an early attempt to quantify embryological research. Based on his own work and that of others, Leuckart concluded that external gill filaments are transitory structures that are common to the embryos of all chondrichthyan fishes. He suggested that passage of the embryos through an external gill filament phase is a type of metamorphosis. In addition, for the first time he describes the yolk stalk appendiculae of the pre-implantation embryos of the bonnethead shark, Sphyrna tiburo. They are vascularized villiform projections that festoon the umbilical cord of this placental species.

Johannes Muller was one of the pivotal figures in mid-nineteenth century biology as a pioneer marine biologist, a founder of comparative physiology, an accomplished embryologist, and a mentor of illustrious students such as Henle, K6lliker, Haeckel, His, Virchow, and Du Bois-Reymond. He helped to lay the foundations for the experimental approach to embryology. Muller's classical paper of 1842 is considered a benchmark of chondrichthyan research. In it, he employed a comparative approach to the study of oviparous and viviparous species. He described and compared the egg cases of sharks, skates, and chimaeras. More importantly, he provided a detailed description of the yolk sac placenta in the smooth dogfish shark, Mustelus canis and the blue shark, Prionace glauca. He gave an in-depth account of the maternal and fetal portions of the placenta and their vascularization. Prior to Mller's definitive account, knowledge of the shark yolk sac placenta had languished since Steno's (1673) rediscovery of it, over 170 years previously. Shortly thereafter, Leydig's (1852) monograph on the histology and embryology of rays and sharks appeared. In it he described five stages in the early development of the spiny dogfish, Squalus acanthias. A detailed description and accurate illustration of the microanatomy of the earliest stage (equivalent to Stage 23 of Ballard et al., 1993) is given. His research on the yolk sac placenta extended that of Miller. The embryonic portion of the placenta of M. laevis at its attachment site is illustrated and a schematic illustration of the histology of the maternal embryonic placental interface is provided. In another advance,

Leydig (1852) described the differentiation of six tissues, viz. neurons, notochord, connective tissue, lens fibres, heart muscle, and striated muscle. In the latter tissue, cell fusion was depicted.

Although the eggs and embryos of viviparous species continued to be used for the study of development, the study of viviparity per se entered a period of stasis that lasted well into the twentieth century. However, some progress was made. Bruch (1860) published the first accurate descriptions and illustrations of the uterine villi of several viviparous rays. The extensive vascularization and glandular organization of these structures were noted. These observations were confirmed and extended by Trois (1876). Wood-Mason and Alcock (1891) discovered that the uterine villi of the butterfly ray (Gymnura) are grossly hypertrophied in one region of the uterus and enter the spiracles of the embryo. This juxtaposition of maternal and embryonic tissues was subsequently termed "a branchial placenta". WoodMason and Alcock coined the term "trophonemata," literally growth threads, to describe the hypertrophied uterine villi. They ascribe a function of nutrient production to the trophonemata. Developing embryos would imbibe the nutrients. Although studies of the yolk sac placenta also lagged, Ercolani's (1879) observations on shark placentae were incorporated into his general classification of the types of vertebrate placentae. Parker (1889) described the fetal membranes of Mustelus antarcticus and provided an analysis of the periembryonic fluid. Alcock (1890) described the histology of the spatulate extensions that occur on the umbilical cord of the yolk sac placenta in the hammerhead shark Sphyrna blochi. He uses, apparently for the first time, the term "appendiculae" to categorize these processes. Their function was enigmatic. Alcock speculates on a possible role as a lymphatic gland or involvement in the "purification" of embryonic blood.

Advances in embryological microtechnique and microscopy that so profoundly altered the study of teleostean embryology had an even more pronounced effect on the study of chondrichthyans because of their large, opaque, meroblastic eggs. Research shifted to the histological and cellular levels. The consequences are especially apparent in research on early development, i.e., fertilization through early development, as well as organogenesis and cell differentiation.

Although the blastoderm and neurula stages had been recognized by Lorenzini (1678), little progress had been made in the intervening years in defining the blastula and gastrula stages. Coste (1850) appears to be the first to describe but not illustrate cleavage. His report was extended by Gerbe (1872) who illustrated surface views of stages from the first cleavage through a late blastula and compared them to the cleavage stages of birds and reptiles. These observations were confirmed by Balfour (1878) who also included illustrations of sectioned blastulae. The definitive study of early development from fertilization through the early gastrula was conducted by Ruckert (1899). Surface views of the blastoderm were accompanied by sections through it.

In 1878, Balfour published his monograph, On the Development of Elasmobranch Fishes. This work raised the study of chondrichthyan development to heretofore unattained heights by establishing new standards of excellence for accuracy and comprehensive description. It has been categorized as "a famous landmark in the history of vertebrate development". Balfour explored all aspects of elasmobranch embryology from early cleavage through mid-late development and established a staged series of embryos, mostly sharks, that was employed in research for many years. Illustrations of sectioned material were used extensively. Balfour described the organization of the mature egg and cleavage of the zygote, commenting on yolk platelets, the germinal vesicle and its fate, the process of cleavage, nuclear division, the origin of yolk nuclei, and asymmetry of the blastoderm. An indepth account of gastrulation, germ layer formation, and the general details of embryonic development is presented. Detailed descriptions of the development of the major organ systems included: (1) brain, spinal and sympathetic nerves; (2) head and associated structures; (3) digestive tube and associated organs; (4) heart and vascular system; and (5) gonads, kidney, Wolffian and Mllerian ducts and their derivatives. In passing, it should be noted that Balfour conducted much of his research at the Naples Zoological Station. His research was facilitated by the director, Anton Dohrn, who was pursuing a major research program in evolutionary physiological anatomy. This program included studies of shark embryology by Dohrn and his associates in order to resolve Dohrn's ideas about vertebrate origins.

Observations by Ziegler and Ziegler (1892) on the development of the electric ray Torpedo continued Balfour's pioneering approach. Their work is characterized by the use of serial sections and photographs of surface views of model reconstructions of selected embryonic stages. They investigated the equivalent of Balfour stages B to K, and considerably advanced descriptive knowledge of gastrulation, formation of the medullary plate, closure of the medullary plate and early embryo formation, and the organization of gill slit-tail bud stage embryos. Balfour had made major advances in the study of chondrichthyan development and the introduction of a series of normal stages was useful. However, there were defects in his normal series. The series was intended to illustrate development in Scyliorhinus, but it was necessary to use embryos of the shark Galeus and the ray Torpedo to fill in gaps. Moreover, there was no time base to the series. As part of Keibel's series of monographs on the normal stages of development of vertebrates, Scammon (1911) published his monograph on the spiny dogfish shark, Squalus acanthias. It is a reasonably complete survey of the development of a single elasmobranch species. Ballard et aL (1993, p. 334) refer to Scammon's monographs as "hitherto unrivaled in its completeness, records in handsome drawings, wax reconstructions and serial sections the anatomy of dozens of randomly collected but closely spaced specimens." Scammon (1911) also included a bibliography arranged by subject of the important literature on chondrichthyan development. The appearance of Scammon's monograph represents the apex of nineteenth century studies of chondrichthyans.

A major conceptual difficulty encountered in the study of chondrichthyan development is the lack of correspondence of the processes of gastrulation and embryo formation with those of other vertebrates with meroblastic eggs, in particular, teleosts and reptiles and birds. Suffice to say that the posterior rim of the chondrichthyan blastoderm thickens to form a medial embryonic shield with two lateral crescent-like arms. The arms eventually close to form the embryonic axis. The embryo-forming region overhangs the yolk mass. Initially, the anterior end of the embryonic axis is a fixed point and lengthening of the axis is accomplished by posterior growth brought about by the joining together of the right and left arms of the shield Although considerable effort has been invested in the descriptive and experimental study of the problem, according to Ballard et al. (1993, p. 328), "No consensus has been reached as to how the morphogenetic cell movements are taking place in this or any other elasmobranch fish."

Another area of inquiry in which the advances in embryological microtechnique provided spectacular insights was the study of oogenesis. Although lampbrush chromosomes were first seen by Flemming in 1878, it was Ruckert (1892a) who first gave evidence that these structures are chromosomes. He investigated the fate of the chromosomal material during the growth of the oocyte nucleus in the ovaries of three elasmobranchs, Scyllium (=Scyliorhinus), Torpedo, and Pristiurus (=Galeus). Ruckert made three major contributions. He was the first to devise a method of isolating intact germinal vesicles and staining lampbrush chromosomes in situ. Secondly, he described the characteristic structure of the lampbrush chromosomes, especially the lateral loops, and was the first to use the term "lampbrush." Finally, by examining stages in oogenesis, he described the genesis of lampbrush chromosomes during early stages of oocyte growth (18 wm to 2 mm), their unique structural configuration at full development (2-3 mm), and the retraction of the lateral loops and contraction of the chromosome axis as these chromosomes give rise to normal condensed meiotic bivalents during later stages (3-13 mm) of oogenesis. Rickert's findings were fully confirmed by Marechal (1907) in an elegantly illustrated study of two sharks and two teleosts.

Within the context of the reproductive system, some other notable contributions should be mentioned. Woods (1902) carried out a quantitative study in which he discovered that the germ cells of the spiny dogfish shark Squalus acanthias originate in endoderm associated with the yolk sac and migrate apparently by amoeboid movement to the germinal ridge of the developing gonad. Borcea's (1905) classical study of the urogenital system contains the definitive statement on the structure of the nidamental (=shell) gland of the oviduct and its role in the secretion and morphogenesis of the egg case.

By the end of the nineteenth century, the descriptive aspects of chondrichthyan development were well known. The state of knowledge compared favorably with what was available for the more intensively studied domestic fowl and selected amphibians. Building on this foundation, future investigators would be able to formulate a series of questions based upon a functional and analytical approach to the study of development.

REPRODUCTIVE SYSTEM

Organs

Fish reproductive organs include testes and ovaries. In most fish species, gonads are paired organs of similar size, which can be partially or totally fused. There may also be a range of secondary reproductive organs that help in increasing a fish's fitness.

In terms of spermatogonia distribution, the structure of teleosts testes has two types: in the most common, spermatogonia occur all along the seminiferous tubules, while in Atherinomorph fishes they are confined to the distal portion of these structures. Fishes can present cystic or semi-cystic spermatogenesis in relation to the phase of release of germ cells in cysts to the seminiferous tubules lumen.

Fish ovaries may be of three types: gymnovarian, secondary gymnovarian or cystovarian. In the first type, the oocytes are released directly into the coelomic cavity and then enter the ostium, then through the oviduct and are eliminated. Secondary gymnovarian ovaries shed ova into the coelom and then they go directly into the oviduct.

In the third type, the oocytes are conveyed to the exterior through the oviduct. Gymnovaries are the primitive condition found in lungfishes, sturgeons, and bowfins. Cystovaries are the condition that characterizes most of the teleosts, where the ovary lumen has continuity with the oviduct. Secondary gymnovaries are found in salmonids and a few other teleosts.

Oogonia development in teleosts fish varies according to the group, and the determination of oogenesis dynamics allows the understanding of maturation and fertilization processes. Changes in the nucleus, ooplasm, and the surrounding layers characterize the oocyte maturation process.

Postovulatory follicles are structures formed after oocyte release; they do not have endocrine function, present a wide irregular lumen, and are rapidly reabosrbed in a process involving the apoptosis of follicular cells. A degenerative process called follicular atresia reabsorbs vitellogenic oocytes not spawned. This process can also occur, but less frequently, in oocytes in other development stages.

Some fish are hermaphrodites, having testes and ovaries either at different phases in their life cycle or, like hamlets, can be simultaneously male and female.

Reproductive Method

Over 97% of all known fishes are oviparous, that is, the eggs develop outside the mother's body. Examples of oviparous fishes include salmon, goldfish, cichlids, tuna, and eels. In the majority of these species, fertilisation takes place outside the mother's body, with the male and female fish shedding their gametes into the surrounding water.

However, a few oviparous fishes practise internal fertilisation, with the male using some sort of intromittent organ to deliver sperm into the genital opening of the female, most notably the oviparous sharks, such as the horn shark, and oviparous rays, such as skates. In these cases, the male is equipped with a pair of modified pelvic fins known as claspers.

The newly-hatched young of oviparous fish are called larvae. They are usually poorly formed, carry a large yolk sac (from which they gain their nutrition) and are very different in appearance to juvenile and adult specimens of their species.

The larval period in oviparous fish is relatively short however (usually only several weeks), and larvae rapidly grow and change appearance and structure (a process termed metamorphosis) to resemble juveniles of their species. During this transition larvae use up their yolk sac and must switch from yolk sac nutrition to feeding on zooplankton prey, a process which is dependent on zooplankton prey densities and causes many mortalities in larvae.

Ovoviviparous fish are ones in which the eggs develop inside the mother's body after internal fertilization but receive little or no nutrition from the mother, depending instead on the yolk. Each embryo develops in its own egg. Familiar examples of ovoviviparous fishes include guppies, angel sharks, and coelacanths.

Some species of fish are viviparous. In such species the mother retains the eggs, as in ovoviviparous fishes, but the embryos receive nutrition from the mother in a variety of different ways. Typically, viviparous fishes have a structure analogous to the placenta seen in mammals connecting the mother's blood supply with the that of the embryo. Examples of viviparous fishes of this type include the surf-perches, splitfins, and lemon shark.

The embryos of some viviparous fishes exhibit a behaviour known as oophagy where the developing embryos eat eggs produced by the mother. This has been observed primarily among sharks, such as the shortfin mako and porbeagle, but is known for a few bony fish as well, such as the halfbeak *Nomorhamphus ebrardtii*. Intrauterine cannibalism is an even more unusual mode of vivipary, where the largest embryos in the uterus will eat their weaker and smaller siblings. This behaviour is also most commonly found among sharks, such as the grey nurse shark, but has also been reported for *Nomorhamphus ebrardtii*.

Aquarists commonly refer to ovoviviparous and viviparous fishes as livebearers.

CLASSIFICATION

Fish are a paraphyletic group: that is, any clade containing all fish also contains the tetrapods, which are not fish. For this reason, groups such as the "Class Pisces" seen in older reference works are no longer used in formal classifications.

Fish are classified into the following major groups:

- Subclass Pteraspidomorphi (early jawless fish)
- Class Thelodonti
- Class Anaspida
- (unranked) Cephalaspidomorphi (early jawless fish);
  - (unranked) Hyperoartia
  - Petromyzontidae (lampreys)
  - Class Galeaspida
  - Class Pituriaspida
  - Class Osteostraci.
- Infraphylum Gnathostomata (jawed vertebrates);
  - Class Placodermi (armoured fishes, extinct)
  - Class Chondrichthyes (cartilaginous fish)
  - Class Acanthodii (spiny sharks, extinct)
  - Superclass Osteichthyes (bony fish)
  - Class Actinopterygii (ray-finned fish)
  - Class Sarcopterygii (lobe-finned fish)
  - Subclass Coelacanthimorpha (coelacanths)
  - Subclass Dipnoi (lungfish).

Some palaeontologists consider that Conodonta are chordates, and so regard them as primitive fish. For a fuller treatment of classification.

The various fish groups taken together account for more than half of the known vertebrates. There are almost 28,000 known extant species of fish, of which almost 27,000 are bony fish, with the remainder being about 970 sharks, rays, and chimeras and about 108 hagfishes and lampreys. A third of all of these species are contained within the nine largest families; from largest to smallest, these families are Cyprinidae, Gobiidae, Cichlidae, Characidae, Loricariidae, Balitoridae, Serranidae, Labridae, and Scorpaenidae. On the other hand, about 64 families are monotypic, containing only one species. It is predicted that the eventual number of total extant species will be at least 32,500.

ANATOMY

Digestive System

The advent of jaws allowed fish to eat a much wider variety of food, including plants and other organisms. In fish, food is ingested through the mouth and then broken down in the esophagus. When it enters the stomach, the food is further broken down and, in many fish, further processed in fingerlike pouches called pyloric caeca. The pyloric caeca secrete digestive enzymes and absorb nutrients from the digested food. Organs such as the liver and pancreas add enzymes and various digestive chemicals as the food moves through the digestive tract. The intestine completes the process of digestion and nutrient absorption.

Respiratory System

Most fish exchange gases by using gills that are located on either side of the pharynx. Gills are made up of threadlike structures called filaments. Each filament contains a network of capillaries that allow a large surface area for the exchange of oxygen and carbon dioxide. Fish exchange gases by pulling oxygen-rich water through their mouths and pumping it over their gill filaments. The blood in the capillaries flows in the opposite direction to the water, causing counter current exchange. They then push the oxygen-poor water out through openings in the sides of the pharynx. Some fishes, like sharks and lampreys, possess multiple gill openings. However, most fishes have a single gill opening on each side of the body. This opening is hidden beneath a protective bony cover called an operculum.

Juvenile bichirs have external gills, a very primitive feature that they hold in common with larval amphibians.

Many fish can breathe air. The mechanisms for doing so are varied. The skin of anguillid eels may be used to absorb oxygen. The buccal cavity of the electric eel may be used to breathe air. Catfishes of the families Loricariidae, Callichthyidae, and Scoloplacidae are able to absorb air through their digestive tracts. Lungfish and bichirs have paired lungs similar to those of tetrapods and must rise to the surface of the water to gulp fresh air in through the mouth and pass spent air out through the gills. Gar and bowfin have a vascularised swim bladder that is used in the same way. Loaches, trahiras, and many catfish breathe by passing air through the gut. Mudskippers breathe by absorbing oxygen across the skin (similar to what frogs do). A number of fishes have evolved so-called accessory breathing organs that are used to extract oxygen from the air. Labyrinth fish (such as gouramis and bettas) have a labyrinth organ above the gills that performs this function. A few other fish have structures more or less resembling labyrinth organs in form and function, most notably snakeheads, pikeheads, and the Clariidae family of catfish.

Being able to breathe air is primarily of use to fish that inhabit shallow, seasonally variable waters where the oxygen concentration in the water may decline at certain times of the year. At such times, fishes dependent solely on the oxygen in the water, such as perch and cichlids, will quickly suffocate, but air-breathing fish can survive for much longer, in some cases in water that is little more than wet mud. At the most extreme, some of these air-breathing fish are able to survive in damp burrows for weeks after the water has otherwise completely dried up, entering a state of aestivation until the water returns.

Fish can be divided into obligate air breathers and facultative air breathers. Obligate air breathers, such as the African lungfish, *must* breathe air periodically or they will suffocate.

Facultative air breathers, such as the catfish *Hypostomus plecostomus*, will only breathe air if they need to and will otherwise rely solely on their gills for oxygen if conditions are favourable. Most air breathing fish are not obligate air breathers, as there is an energetic cost in rising to the surface and a fitness cost of being exposed to surface predators.

Circulatory System

Fish have a closed circulatory system with a heart that pumps the blood in a single loop throughout the body. The blood goes from the heart to gills, from the gills to the rest of the body, and then back to the heart. In most fish, the heart consists of four parts: the sinus venosus, the atrium, the ventricle, and the bulbus arteriosus.

Despite consisting of four parts, the fish heart is still a two-chambered heart. The sinus venosus is a thin-walled sac that collects blood from the fish's veins before allowing it to flow to the atrium, which is a large muscular chamber. The atrium serves as a one-way compartment for blood to flow into the ventricle. The ventricle is a thick-walled, muscular chamber and it does the actual pumping for the heart. It pumps blood to a large tube called the bulbus arteriosus. At the front end, the bulbus arteriosus connects to a large blood vessel called the aorta, through which blood flows to the fish's gills.

Excretory System

As with many aquatic animals, most fish release their nitrogenous wastes as ammonia. Some of the wastes diffuse through the gills into the surrounding water. Others are removed by the kidneys, excretory organs that filter wastes from the blood. Kidneys help fishes control the amount of ammonia in their bodies. Saltwater fish tend to lose water because of osmosis.

In saltwater fish, the kidneys concentrate wastes and return as much water as possible back to the body. The reverse happens in freshwater fish: they tend to gain water continuously. The kidneys of freshwater fish are specially adapted to pump out large amounts of dilute urine. Some fish have specially adapted kidneys that change their function, allowing them to move from freshwater to saltwater.

SENSORY AND NERVOUS SYSTEM

Central Nervous System

Fish typically have quite small brains relative to body size when compared with other vertebrates, typically one-fifteenth the mass of the brain from a similarly sized bird or mammal. However, some fish have relatively large brains, most notably mormyrids and sharks, which have brains of about as massive relative to body weight as birds and marsupials.

The brain is divided into several regions. At the front are the olfactory lobes, a pair of structure the receive and process signals from the nostrils via the two olfactory nerves.

The olfactory lobes are very large in fishes that hunt primarily by smell, such as hagfish, sharks, and catfish. Behind the olfactory lobes is the two-lobed telencephalon, the equivalent structure to the cerebrum in higher vertebrates. In fishes the telencephalon is concerned mostly with olfaction. Together these structures form the forebrain.

Connecting the forebrain to the *midbrain* is the diencephalon. The diencephalon performs a number of functions associated with hormones and homeostasis. The pineal body lies just above the diencephalon. This structure performs many different functions including detecting light, maintaining circadian rhythms, and controlling colour changes.

The midbrain or mesencephalon contains the two optic lobes. These are very large in species that hunt by sight, such as rainbow trout and cichlids.

The hindbrain or metencephalon is particularly involved in swimming and balance. The cerebellum is a single-lobed structure that is usually very large, typically the biggest part of the brain. Hagfish and lampreys have relatively small cerebellums, but at the other extreme the cerebellums of mormyrids are massively developed and apparently involved in their electrical sense.

The brain stem or myelencephalon is the most posterior part of the brain. As well as controlling the functions of some of the muscles and body organs, in bony fish at least the brain stem is also concerned with respiration and osmoregulation.

Sense Organs

Most fish possess highly developed sense organs. Nearly all daylight fish have well-developed eyes that have colour vision that is at least as good as a human's. Many fish also have specialized cells known as chemoreceptors that are responsible for extraordinary senses of taste and smell.

Although they have ears in their heads, many fish may not hear sounds very well. However, most fishes have sensitive receptors that form the lateral line system. The lateral line system allows for many fish to detect gentle currents and vibrations, as well as to sense the motion of other nearby fish and prey. Some fishes such as catfishes and sharks, have organs that detect low levels electric current. Other fish, like the electric eel, can produce their own electricity.

Pain Reception in Fish

In 2003, Scottish scientists at the University of Edinburgh performing research on rainbow trout concluded that fish exhibit behaviours often associated with pain. Professor James D. Rose of the University of Wyoming critiqued the study, claiming it was flawed. Rose had published his own study a year earlier arguing that fish cannot feel pain as they lack the appropriate neocortex in the brain.

Muscular System

Most fish move by contracting paired sets of muscles on either side of the backbone alternately. These contractions form S-shaped curves that move down the body of the fish. As each curve reaches the back fin, backward force is created.

This backward force, in conjunction with the fins, moves the fish forward. The fish's fins are used like an airplane's stabilizers. Fins also increase the surface area of the tail, allowing for an extra boost in speed. The streamlined body of the fish decreases the amount of friction as they move through water. Since body tissue is more dense than water, fish must compensate for the difference or they will sink. Many bony fishes have an internal organ called a swim bladder that adjusts their buoyancy through manipulation of gases.

IMMUNE SYSTEM

Types of immune organs vary between different types of fish. In the jawless fish (lampreys and hagfishes), true lymphoid organs are absent. Instead, these fish rely on regions of lymphoid tissue within other organs to produce their immune cells. For example, erythrocytes, macrophages and plasma cells are produced in the anterior kidney (or pronephros) and some areas of the gut (where granulocytes mature) resemble primitive bone marrow in hagfish.

Cartilaginous fish (sharks and rays) have a more advanced immune system than the jawless fish. They have three specialized organs that are unique to chondrichthyes; the epigonal organs (lymphoid tissue similar to bone marrow of mammals) that surround the gonads, the Leydig's organ within the walls of their esophagus, and a spiral valve in their intestine. All these organs house typical immune cells (granulocytes, lymphocytes and plasma cells).

They also possess an identifiable thymus and a well-developed spleen (their most important immune organ) where various lymphocytes, plasma cells and macrophages develop and are stored. Chondrostean fish (sturgeons, paddlefish and birchirs) possess a major site for the production of granulocytes within a mass that is associated with the meninges (membranes surrounding the central nervous system) and their heart is frequently covered with tissue that contains lymphocytes, reticular cells and a small number of macrophages. The chondrostean kidney is an important hemopoietic organ; where erythrocytes, granulocytes, lymphocytes and macrophages develop.

Like chondrostean fish, the major immune tissues of bony fish (or teleostei) include the kidney (especially the anterior kidney), where many different immune cells are housed. In addition, teleost fish possess a thymus, spleen and scattered immune areas within mucosal tissues (e.g. in the skin, gills, gut and gonads). Much like the mammalian immune system, teleost erythrocytes, neutrophils and granulocytes are believed to reside in the spleen whereas lymphocytes are the major cell type found in the thymus.

Recently, a lymphatic system similar to that described in mammals was described in one species of teleost fish, the zebrafish. Although not confirmed as yet, this system presumably will be where naive (unstimulated) T cells will accumulate while waiting to encounter an antigen.

Evolution

The early fossil record on fish is not very clear. It became a dominant form of sea life and eventually branched to create land vertebrates.

The proliferation was apparently due to the formation of the hinged jaw because jawless fish left very few descendants. Lampreys may be a rough representative of pre-jawed fish. The first jaws are found in Placodermi fossils. It is unclear if the advantage of a hinged jaw is greater biting force, respiratory-related, or a combination.

Some speculate that fish may have evolved from a creature similar to a coral-like Sea squirt, whose larvae resemble primitive fish in some key ways. The first ancestors of fish may have kept the larval form into adulthood (as some sea squirts do today), although maybe the reverse of this is case. Candidates for early fish include Agnatha such as Haikouichthys, Myllokunmingia, Pikaia, and Conodonts.

Homeothermy

Although most fish are exclusively aquatic and ectothermic, there are exceptions to both cases.

Fish from a number of different groups have evolved the capacity to live out of the water for extended periods of time. Of these amphibious fish, some such as the mudskipper can live and move about on land for up to several days.

Also, certain species of fish maintain elevated body temperatures to varying degrees. Endothermic teleosts (bony fishes) are all in the suborder Scombroidei and include the billfishes, tunas, and one species of "primitive" mackerel (*Gasterochisma melampus*).

All sharks in the family Lamnidae – shortfin mako, long fin mako, white, porbeagle, and salmon shark – are known to have the capacity for endothermy, and evidence suggests the trait exists in family Alopiidae (thresher sharks). The degree of endothermy varies from the billfish, which warm only their eyes and brain, to bluefin tuna and porbeagle sharks who maintain body temperatures elevated in excess of 20 °C above ambient water temperatures. Endothermy, though metabolically costly, is thought to provide advantages such as increased contractile force of muscles, higher rates of central nervous system processing, and higher rates of digestion.

Diseases

Like other animals, fish can suffer from a wide variety of diseases and parasites. To prevent disease they have a variety of non-specific defences and specific defences. Non-specific defences include the skin and scales, as well as the mucus layer secreted by the epidermis that traps microorganisms and inhibits their growth. Should pathogens breach these defences, fish can develop an inflammatory response that increases the flow of blood to the infected region and delivers the white blood cells that will attempt to destroy the pathogens.

Specific defences are specialised responses to particular pathogens recognised by the fish's body, in other words, an immune response. In recent years, vaccines have become widely used in aquaculture and also with ornamental fish, for example the vaccines for furunculosis in farmed salmon and koi herpes virus in koi. Some fish will also take advantage of cleaner fish for removal of external parasites.

The best known of these are the Bluestreak cleaner wrasses of the genus *Labroides* found on coral reefs in the Indian Ocean and Pacific Ocean. These small fish maintain so-called "cleaning stations" where other fish, known as hosts, will congregate and perform specific movements to attract the attention of the cleaner fish. Cleaning behaviours have been observed in a number of other fish groups, including an interesting case between two cichlids of the same genus, *Etroplus maculatus*, the cleaner fish, and the much larger *Etroplus suratensis*, the host.

EVOLUTIONARY IMPACT OF FACULTATIVE PARTHENOGENESIS

It is a widely accepted assumption that in animals parthenogenesis is a secondary mode of reproduction and that all clonally reproducing animals have sexually reproducing ancestors. This hypothesis is based on the phylogenetic distribution of clonal reproduction in animals that can always be found at the tips of tree branches in many different animal groups. Considering the severe genetic disadvantages of clonal reproduction mentioned before, why should a species/individual switch to clonal reproduction? Is clonal reproduction in animals indeed a reproductive error or might there be circumstances that would give an adaptive advantage to parthenogenetically reproducing genotypes? In fact, several evolutionary scenarios are imaginable where the benefits of parthenogenesis (fast reproductive rates, independence from males) might be advantageous. This might be the case e.g. during the colonization of new habitats. The fast reproductive rates of parthenogenesis could be a great advantage for the exploitation and monopolization of new resources and suppress slower competitors (monopolization hypothesis). Also, if population sizes are still low and males are scarce, parthenogenetic females would clearly have an advantage compared to sexual females. In fact, parthenogens are often observed in disturbed areas and on the edges of species distributions (geographical parthenogenesis). While some researchers argue that parthenogenetic species are inferior to sexually reproducing species and are therefore pushed to distribution edges, others claim that parthenogenetic species might simply be better suited to pioneer new habitats. The faster reproductive rate of parthenogens could also be an advantage in environments with seasonal changes in food availability. During times with high food abundance parthenogens should have an advantage in using the available food (bottom-up regulation), while at other times variable offspring might be advantageous to escape mortality (top-down regulation). Scarcity of males might also be a problem independently of the colonization of new habitats. Low population sizes and a lack of males might occur e.g. in seasonal climates with harsh winter conditions or when very scattered resources are explored. From these scenarios it becomes clear that even though sexual reproduction is the original mode of reproduction in vertebrates and clonal reproduction has severe genetic costs, there are conditions that give parthenogenetic reproduction an evolutionary adaptive advantage.

Nevertheless, facultative parthenogenesis that might combine the advantages of sexual and clonal reproduction is very rare in vertebrates. In vertebrates, switches to parthenogenetic reproduction seem to be a reaction to scarcity of males and not so much to food availability and habitat monopolization. Even though it has, so far, only been observed in captive females that were kept isolated from males for very long time periods, all case reports agree that facultative parthenogenesis could take place in natural populations. Facultative parthenogenesis in natural populations might have been missed due to the fact that it is most likely hard to recognize but also because until recently it received little scientific attention. While all-female reproduction in natural populations was recognized early on, cases of facultative parthenogenesis in bisexual species may have simply been overlooked or may have been misclassified, e.g. as sperm storage events. However, an increasing amount of events is being reported and comes from nearly all vertebrate groups.

Interestingly, facultative parthenogens often use automixis to restore the ploidy of their oocytes, therefore producing genotypically diverse offspring despite reproducing parthenogenetically. In the ZW sex-determining system automixis and to be more precise terminal fusion leading to homozygosity has a very interesting effect: parthenogenetically reproducing females which are the heterozygous sex at the sex chromosome (ZW) can have homozygous and therefore male offspring. In terminal fusion automixis WW and ZZ offspring will be produced. While WW is not viable. Parthenogenetically derived males have been found in chicken, turkey, lizards and snakes and have been reported to successfully reproduce with females after maturation. This means that females that have to reproduce by facultative parthenogenesis because males are scarce could produce offspring with both sexes and in this way revert to sexual reproduction within a single generation.

This mechanism, that directly counteracts the scarcity of males, might be beneficial even if the individual reproductive output is low due to mutationally burdened offspring. It might be one explanation for the observation that facultative parthenogenesis is more often observed in species with the ZW (reptiles, birds) than the XY (mammals) sex-determination system, even though alternative, e.g. phylogenetic, explanations for this observation cannot be disregarded. It also seems striking that, even though islands do not harbour more parthenogenetic species than the mainland, they are very often predominantly inhabited by birds and reptiles, the two vertebrate groups where ZW sex determination is a common mode of reproduction. This might of course in large parts be due to the high mobility of these vertebrate groups and to sperm storage but could, to a lesser degree, also be explained by a genetic predisposition to facultative parthenogenesis in these groups. Population genetic field studies focusing on sex-determination system and genomewide male homozygosity in island populations would be needed to determine the roles of the sex-determination system and facultative parthenogenesis in colonization processes.

While in vertebrates facultative parthenogenesis has long been viewed as a reproductive mistake brought about by captivity and unnaturally long periods of isolation from males, it might in fact be a great evolutionary chance especially for colonization and might explain the biodiversity patterns found on island habitats. Future studies are needed that focus on the detection of facultative parthenogenesis in natural populations and detailed studies of island biogeography might allow for a less speculative and more founded look at the impact of sex determination on the potential for facultative parthenogenesis. Recent advances in molecular methods have provided tools for fast and easy analyses of reproductive mechanisms even in natural populations. Microsatellites in particular seem to be very useful tools for the analysis of unreduced oocyte formation and are routinely used in the more recent studies. Facultative parthenogenesis is already discussed as a conservation issue in zoological facilities but, as it is reported from an increasing number of species in all vertebrate groups, urgently needs to be investigated in natural populations.

Even though facultative parthenogenesis is very rare in vertebrates and has so far only been observed in captive animals, it has been reported in an increasing number of species in recent years. All reports from captivity agree that although facultative parthenogenesis has so far only been observed in artificial situations, it might occur in natural populations. Facultative parthenogenesis has been reported quite frequently in birds and reptiles where it can even lead to viable and fertile offspring. Parthenogenetic development has also been reported in mammals, but viable offspring here seems to be prevented by genetic imprinting. Due to the lack of reports from natural populations, the evolutionary impact of facultative parthenogenetic reproduction in vertebrates is so far impossible to determine. Theoretical considerations call for a significant impact under special circumstances, e.g. the colonization of new habitats (islands). The evolutionary impact must be correlated to the ploidy restoration mechanism used during oocyte formation and to the sex-determination system of the species. With the molecular techniques available now, more cases of facultative parthenogenesis will be identified in vertebrates, and we will get deeper insights into mechanisms allowing and counteracting the formation and development of unreduced eggs. Due to its potential evolutionary impact as well as medical importance, facultative parthenogenesis in vertebrates gains more and more scientific interest and will link different fields as ecology, evolution and conservation biology with medical research.

8

Genetic Structure of Fish

Biochemical genetic markers have been increasingly used for inferences on the population genetic structure of various fish, but little attention has been paid to relative differences in the distribution of variability within species in different groups of fish. Worldwide, fisheries provide a source of employment for millions of people and a source of protein for billions. The management of fisheries is therefore a critical activity. Unfortunately, the realisation that fish are a limited resource was slow to reach acceptance, and there is not much evidence that attempts to manage fisheries have led to sustainability. In 1865, a group led by T.H. Huxley published the Sea Fisheries Commission Report into British fisheries, the first of its kind, in which it was recommended that there should be as few restrictions as possible on inshore and offshore fisheries and the fishing gear used. In effect, the conclusion of the report was that the supply of fish was inexhaustible. One hundred years later it was clear that this was patently not the case, and in more recent times many of the world's major fisheries have collapsed – or are on the verge of doing so – through overfishing and poor management.

In order to manage a fishery effectively, a great deal of information is required. We need to know, for example, the size of a population, its habitat and migratory behaviour, its age and size structure, the reproductive pattern of the species, the natural mortality rate, the rate at which fish are removed by fishing, and so on. Most of these factors can be, and are, determined from fishery statistics based on detailed surveys and analyses of landings. However, a critical requirement is to know whether the fish species exists as a single genetic unit or as a series of relatively genetically distinct groups. Is the species genetically homogeneous (panmictic) or is it genetically heterogeneous? If there are local genetically distinct groups, management strategies will need to be adapted to take this into account.

This substructuring into different, relatively genetically discrete groups may take various forms. For example, the migratory salmonids return to the rivers in which their parents spawned and this effectively creates genetic groups that are associated with particular rivers or even particular regions of a river. In another example, some lake fish have autumn and spring spawners, and as long as this character is inherited, two genetically distinct groups can evolve. In many marine fish, there are particular sites where spawning aggregations occur and these can lead to genetic differentiation. In the case of sedentary shellfish, dispersal is usually mainly via a larval phase, and hydrographic factors will constrain the direction and distance of dispersal. Shellfish such as flat oysters with a relatively short larval phase or other species with direct development will be more likely to be genetically differentiated than shellfish such as cupped oysters or mussels with an extended larval phase.

What should we call these genetically differentiated groups and what qualifies as 'genetically differentiated'? It is possible to quantify levels of genetic differences using various indices, but for the moment, let us consider the terminology that is available. Partly or strongly genetically differentiated groups within a species have been called (1) 'varieties', 'races', 'breeds' or 'strains' if they are cultivated and (2) 'stocks', 'demes', 'populations', 'subpopulations', 'types' or 'ecotypes' if they are wild. All of these terms have been used in the past to describe genetic differentiation within species, but they have often not been clearly defined.

In the case of fisheries, the word 'stock' has often been defined by fishery managers as a group of fish exploited by a specific method or existing in a particular area. Although this may be convenient for the analysis of landings or catch and effort data, and for the enforcement of management measures such as quotas or net size, it may bear little relation to the true genetic substructuring of the fish species. The important point to realise is that there is a continuum of genetic differentiation. In some species (or in parts of their range) there may be clearly genetically distinct groups, while in others (or in other parts of their range) there will be groups which are virtually indistinguishable genetically. Thus, the use of a single word such as 'stock' to define the units of such substructuring is not often appropriate. Here, we will use the less prescriptive term of 'population' to define an intraspecific (within a species) group of randomly mating individuals which exist (and can therefore be sampled) in a defined geographic position or at a defined time. A key point to note here is that we are assuming that all individuals comprising a population are able to mate randomly with one another, and therefore, there can be no further genetic subdivision within a population. The main types of genetic markers were outlined, and we now need to show how these markers can be used to identify any population substructure within species of fish or shellfish. We will start with markers such as allozymes, microsatellites and single nucleotide polymorphisms (SNPs) that are co-dominant, that is they allow direct identification of homozygotes and heterozygotes. How do we handle this information on genotypes in a population? Firstly, we should note that the process of meiosis breaks up existing genotypes and that new genotypes are reassembled when egg and sperm combine at syngamy. Therefore, although there will be the same range of genotypes in an offspring generation as in a parent generation, the two are not directly linked by inheritance. What is linked by inheritance, however, is the presence of an allele and the frequency of that allele in the population. And it is the allele frequency (or gene frequency) that is used to investigate the structure of populations within species.

How are allele frequencies estimated?

When genotypes at a co-dominant locus are scored from a sample of individuals, the frequencies of the alleles at the locus can easily be calculated from these data. Remember that every diploid individual has pairs of chromosomes, so they must also have pairs of alleles at each locus, that is a homozygote at a locus has two of the same allele and a heterozygote has two different alleles. Although a diploid individual can only exhibit two alleles at a locus, there may be many different alleles at that locus present in the population as a whole. Allele frequencies are the critical starting point for all further analyses of genetics in populations. By convention, the frequencies of alleles at a locus are symbolised using the lower-case letters *p*, *q*, *r*, *s* and the frequency of an allele is given by:

$$2H_o + H_e p = 2N$$

where H_o is the number of homozygotes for that allele, H_e is the number of heterozygotes for the allele and N is the number of individuals scored at the locus.

What is the relationship between alleles and genotypes?

Considering a single locus, it is convenient to envisage a population as a 'pool' of alleles. Haploid gametes are produced by each generation, and each of these gametes contains a single allele from this pool. Let us assume that all adults in the population provide equal numbers of gametes and that each gamete has an equal chance of combining with all other gametes. Once we know the frequencies of the alleles in the population, it is possible to calculate the frequencies with which the genotypes will be present in the offspring from that generation. This principle was first quantified in 1908 by both Godfrey Hardy, a British mathematician, and Wilhelm Weinberg, a German doctor, and is called, unsurprisingly, the Hardy–Weinberg model .

The random mixing of eggs and sperm the proportions of the genotypes will be:

AA AB BB

p^2 $2pq$ q^2

which, in mathematical terms, is equal to $(p + q)^2$. As long as nothing upsets the Hardy–Weinberg model, allele frequencies should remain constant from generation to generation. The frequencies of the genotypes predicted by the model are dependent on the assumptions of equal and random parental contributions and equal and random mixing of gametes. However, if this is not the case, the model will not hold true. Also, we do not usually sample animals at the beginning of their lives; normally fish or shellfish are sampled as adults. In order for the Hardy–Weinberg model to be appropriate in populations of adults, further assumptions need to be made. We must assume that there is no selection that causes one or other genotype to suffer differential mortality.

We must also assume that our sample taken does not include individuals that have migrated into the population from some other population with different genotype frequencies. So both migration and selection are factors that can cause deviation from the genotype frequencies predicted by the Hardy–Weinberg model.

If the number of parents that reproduce in a population is significantly smaller than the whole adult population, allele frequencies in these parents might differ from the whole population and the allele frequencies in the following generation will therefore change.

A further consideration is that very few fish or shellfish species exist as discrete generations, most populations being mixtures of animals from different overlapping generations.

How do allele frequencies change over time?

If a population is in agreement with the Hardy–Weinberg model, then allele frequencies at a locus are not being driven (by selection or migration) to change significantly from generation to generation. However, there is always bound to be some variation in allele frequency from one generation to the next because of the element of chance. Contrary to the idealised situation required to make mathematical predictions, in real life all individuals of all genotypes will not produce exactly the same number of gametes. Neither are all gametes likely to undergo truly random mixing. Therefore, there will be variation from the model in every generation. This is a fact of biological variability, called random genetic drift. In order to quantify this natural variation, allele frequencies can be given a variance as indicated in the formula:

$p(1\text{-}p)$ Variance of frequency of allele = $2N_e$

where p is the frequency of the allele and N_e is the effective population size. The concept of the effective population size is an important one. It is the number of individuals in the population that contribute genetically to the next generation.

It will exclude juveniles, individuals too old to reproduce and those that provide non-contributory gametes. These may be individuals that have infertile gametes, or those whose gametes never come into contact with the gametes of other members of the population due to geographic position or timing of spawning. Thus it can be seen how the effective population size could, in certain circumstances, be very much smaller than the total number of individuals in the population.

If the effective population size (N_e) in the expression above is a very large number, then the allele frequency variance between successive generations will be very small. Conversely, allele frequency variance will be large when N_e is small. This means that natural fluctuations in allele frequency between generations will be much greater in small populations. We shall encounter this concept again when dealing with various aspects of genetics in small populations.

Allele frequencies in a population can therefore change over time owing to random genetic drift. However, they may also change owing to pressure of selection (one genotype or one allele survives better than others at a locus) or to patterns of migration or dispersal that may fluctuate in direction or in strength over time.

How does population structure arise?

The differentiation of a species into genetically different populations is a fundamental part of the process of evolution. For a number of physical or biological reasons, the distribution of a species may become fragmented. For example, during the last ice ages, aquatic species in the temperate regions of the northern hemisphere were driven south and many became restricted to local areas in southern habitats. Where species were fragmented into different areas, many of these populations remained isolated for long periods of time in such refugia. Following the retreat of the last glaciation some 10 000–13 000 years ago, aquatic and terrestrial species went through the process of gradual recolonisation of habitat and this has provided plenty of opportunity for further fragmentation.

Once fragmentation has occurred, allele frequencies at most loci will be subject to random genetic drift in the fragmented populations. As we have seen, random genetic drift produces more rapid allele frequency change in small, compared with large, populations. If only a few individuals are involved initially in founding a population, then allele frequencies in the new population may be very different from the source population, and if numbers remain small, further allele frequency changes may be rapid.

In addition to random genetic drift, localised adaptation will occur involving selection at some loci for particular characteristics and these will cause further differences between populations. These loci under selection, and to a lesser extent those linked to them, can show higher patterns of differentiation between populations than neutral markers and are sometimes called 'outliers'. These processes of random genetic drift and adaptation will tend to increase differences between populations, while migration or larval dispersal between populations will tend to reduce it. So the population differentiation we can identify today in a species is essentially the result of the historical interplay between the environment and the forces of dispersal, adaptation and random genetic drift.

How are genetic markers used to study population structure?

The level of population structure within a species. Most data exist for allozymes, but there are plenty of examples of the use of mitochondrial DNA (mtDNA) variation to look at stock structure, while microsatellites and SNPs have emerged as the markers of choice in the twenty-first century.

In any population study, the ideal first step would be to collect samples of the species across its entire range to estimate genetic differentiation within the species as a whole. In fact, this is seldom done. For reasons of economy or sampling constraints, most genetic studies related to fisheries have tended to focus on limited sampling in specific areas where there has been a commercial or conservation interest.

Depending on the type of marker employed, genotypes (allozymes, microsatellites, SNPs) or haplotypes (mtDNA) are scored for the individuals sampled and the data are analysed in a variety of ways to quantify levels of genetic variation between populations. One commonly applied type of analysis, called *F*-statistics, was developed by Sewall Wright in the first half of the twentieth century. Because the development of mathematical models requires certain concepts or parameters to be fixed at the start, various assumptions had to be made in the development of *F*-statistics and it is important to consider what these were.

First, Wright made the assumption that all populations were of the same size, that is they consisted of approximately the same number of reproductively active individuals. How valid is this assumption? Well, it is easy to imagine situations, particularly in a heavily fished species, where population size could be very different between populations. Also, in sedentary shellfish, the size of populations will be constrained by, among other factors, the availability of suitable habitat. The second assumption was that there was an equal possibility for any population to exchange individuals with any other population. Anyone who has some knowledge of aquatic habitats such as lakes, rivers and oceans will immediately see the difficulty with this assumption, as only in very special circumstances is it likely to hold true. In most cases, movement of larval individuals between populations is likely to be unidirectional rather than multidirectional due to the effects of currents and there are very few fish species where exchanges of juveniles or adult individuals between populations would be expected to be random in both directions. These two theoretical assumptions are a required part of the 'island model' which Wright used as a basis for *F*-statistics. It was further assumed that changes in allele frequency at most loci over time were essentially due to random genetic drift rather than to selection.

Wright's *F*-statistics provides answers to two different questions. The first question is: for the loci scored, are the genotypes in the proportions predicted by the Hardy–Weinberg model? F_{IS} provides a measure of this agreement for a single population and F_{IT} for all the populations combined. F_{IS} and F_{IT} can vary from -1.0 through 0 to 1.0 and exact agreement to the Hardy–Weinberg model equals 0. The second question is: for the loci scored, are the allele frequencies different between various populations? F_{ST} provides a measure of this population differentiation and ranges from 0, where all populations have the same allele frequencies at all loci, to 1.0 where all populations are fixed for different alleles at all loci.

The Japanese geneticist Masatoshi Nei developed G_{ST}, the coefficient of genetic diversity, which is an equivalent index to Wright's F_{ST}. It is calculated slightly differently, but is essentially addressing the same questions. Because these indices represent the genetic differentiation between populations, it is possible to suggest how many individuals might be being exchanged per generation between these populations in order to produce the amount of differentiation observed. Using the formula $m = (1\text{-}F_{ST})/(4F_{ST}\ N_e)$ and taking the example of large effective population sizes of 1 million ($N_e = 10^6$) with a pairwise comparison between populations of $F_{ST} = 0.000025$, we would calculate that about 1% of individuals in each of the population were migrants (m = 0.01) from other populations. Thus $N_e m$, the number of migrants per generation, would be 10 000. It is easy to see that with that high number of effective migrants per generation there is very little possibility of genetic differentiation developing between populations. On the other hand, with a smaller effective population size ($N_e = 10^3$) and an F_{ST} of 0.2, we would estimate a very low proportion of migrants ($m = 0.001$), with an $N_e m$ of 1.0 per generation.

With the exchange of individuals comes the exchange of genes and the greater the exchange, the less will be the genetic differentiation between the populations. Although there is no hard and fast rule, it is accepted that important or significant genetic differentiation is only likely to arise between populations when fewer than one individual per generation, on average, is being exchanged. Again, this is based on the assumption that changes in allele frequencies are based on random genetic drift alone. Obviously, at some loci, there could be significant changes due to selection and such changes would give the false impression that there is less migration between populations than is actually the case.

There are alternative ways of estimating genetic differentiation besides using Wright's F_{ST} or Nei's G_{ST}. It is possible simply to test for the heterogeneity of allele frequencies at each locus across all populations using Contingency Table tests. Traditional testing would involve the use of the $\div^2$ (chi-square), or *G*, test, but nowadays, with the enormous computing power available in desktop PCs, much safer 'exact' tests can be employed and these are used in modern genetic analysis computer packages. What this test tells us is whether there is significant heterogeneity in allele frequencies across all populations. Selective removal of populations that look particularly different, followed by retesting the data, can reveal further detail. Care is needed to avoid the type I statistical errors associated with the use of several tests of the same hypothesis.

Finally, allele frequencies can be used to calculate pairwise genetic differences between populations, or species, using formulae developed by a number of scientists. Nei's genetic identity (*I*) and genetic distance (*D*) are the most commonly used. Once again, a major difficulty faced by geneticists in trying to describe what can be an extremely complex situation by a single statistic is the problem of the neutrality of the gene loci used. On the one hand, genetic variation at many loci could be adaptive such that selection operates and particular alleles or genotypes are favoured in particular situations. Alternatively, allelic variation at most gene loci could be neutral, that is it varies by random genetic drift and is not subject to selection. For most genetic indices, the usual assumption is that all of the alleles at all of the loci included in the determination of the index are neutral. When just a few markers out of a large panel show significantly higher genetic differentiation than all the others, such 'outliers' may indicate that these markers are linked to traits differentially selected in the populations sampled relative to neutral markers. This illustrates the fact that the more markers studied, the more detailed the information obtained regarding the various evolutionary forces shaping differentiation of populations and their evolution.

MITOCHONDRIAL AND CHLOROPLAST DNA

Mitochondrial DNA (mtDNA) insertions into nuclear chromosomes have been documented in a number of eukaryotes. We used fluorescence in situ hybridization (FISH) to examine the variation of mtDNA insertions in maize. So far we have considered the DNA present in the nucleus, which is organised into chromosomes. There is actually more DNA in the cell – extra-chromosomal genes, contained within energy-generating organelles, mitochondria, of which there may be several hundred in each cell. In plants, the photosynthetic organelles – chloroplasts – also contain DNA. Animal mitochondrial DNA (mtDNA) is normally present as a circular molecule of around 16 kb in length, and there are around ten copies of the DNA in each mitochondrion in humans. Unlike the chromosomal DNA, there is no meiosis and replication appears to be a simple copying process, though recent research does point to there being some form of recombination during mtDNA replication.

Because there are large numbers of mitochondria in an egg, but very few in a spermatozoon, it is hardly surprising to find that the mtDNA present in a sexually reproduced offspring is usually inherited entirely from its mother (strict maternal inheritance). This maternal-only inheritance of mtDNA is the situation in almost all animals.

However, one exception to this rule occurs in an important aquaculture species, the mussel *Mytilus* spp., which has a form of bi-parental inheritance of mtDNA. Females have an F-type of mtDNA in every body cell, while males have both the F-type and an M-type mtDNA in most cells of the body. The M-type is highly concentrated in the male gonad and is thought to be the only mtDNA present in the spermatozoa. These M-type mtDNA molecules present in a spermatozoon enter the egg, and, in some way which is not yet fully understood, the M-type remains in the egg after fertilisation and is eliminated in individuals destined to become female, but retained and preferentially replicated in individuals destined to become males. This unusual arrangement has been named 'doubly uniparental inheritance'. Doubly uniparental inheritance has recently been detected in other bivalves besides mussels such as the manila clam (*Tapes philippinarum*) and may be more widespread than currently thought.

The complete sequence of the mitochondrial genome is now known for quite a number of vertebrates and invertebrates, and the order of the genes within the circular genome is different in every phylum so far studied. Because fish mtDNA has been extensively used in phylogenetic studies, we will use this molecule as an example. The mitochondrial genome of fish contains 13 genes coding for proteins, 2 genes coding for rRNA (the small 12S and the large 16S rRNA), 22 genes coding for tRNA molecules and 1 non-coding section of DNA which acts as the initiation site for mtDNA replication and RNA transcription. This is called the control region.

In contrast to the nuclear genome, the mitochondrial genes of animals are very efficient and have no introns. In addition, there is virtually no 'junk

DNA' or repetitive sequences in the mitochondrial genome, although the control region does often vary in length due to tandem repeats. Exceptions to this general rule are the scallops, many species of which exhibit several large (up to 1.4 kb) repeated sequences within the mtDNA genome which can consequently extend to beyond 30 kb in length.

For reasons which are not fully understood, the rate of mutation in animal mtDNA is higher than that in the nuclear DNA (about 5–10 times higher). This means that the rate of evolution is greater in mtDNA than in nuclear DNA, and this feature is of importance to us when we are looking for genetic markers which will reflect changes in the more recent past.

Chloroplasts of photosynthetic organisms also contain DNA molecules, having originated from endosymbiotic cyanobacteria. Each chloroplast can contain up to 100 copies of this small genome composed of circular molecules. It is commonly maternally inherited. The chloroplast genome was the first plant genome characterised because of its small size and ease of isolation. The complete chloroplast genome sequences of a number of plants and algae have been determined, revealing massive transfer of genes from ancestral organelles to the nucleus. This might explain why all chloroplast functions require imported proteins encoded by the nuclear genome.

HOW CAN GENETIC VARIATION BE MEASURED?

Genetic variation can be measured and quantified at several levels. Firstly, the precise sequence of a DNA fragment, and how it varies between individuals, can be determined. Secondly, differences between sizes of DNA fragments can be identified. At the next level, we can consider protein differences that result from DNA coding sequence variation. Finally, it is sometimes possible to identify phenotypic differences that are the product of genetic variation at just one or two loci or for traits determined by a much larger number of genes. It is of interest to note that the actual chronology of the use of these various levels of information is precisely the reverse of the way that they are introduced in this chapter. The first markers of genetic variation to be described were exclusively phenotypic, and even though the structure of DNA was discovered in 1953, they were still the only markers used until the mid-1960s when variations in proteins (allozymes) were identified by electrophoresis. Mitochondrial DNA markers based on fragment lengths were developed in the 1980s, but the huge leap forward came in the 1990s with the introduction of the polymerase chain reaction that allowed cheap and rapid amplification of fragments of DNA that could be sequenced. There has been a very rapid rate of change and improvement in DNA technologies since the turn of the century, enabling the accumulation of huge databases of DNA sequences, the whole genomes of some organisms (including humans), hundreds of thousands of functional DNA sequences for many species and innumerable protein sequences.

POTENTIAL PROBLEMS WITH ALLOZYMES AND CODING MARKERS

There is a risk that allozyme markers could fail to detect genetic isolation because, as enzymes, they can be vulnerable to selection operating at the biochemical level, either directly on them or on the biochemical pathways within which they operate. This can be also the case for SNP markers located in exons or any marker closely linked to another one that is subject to selection.

If the selection is such that it is in the same direction in two genetically isolated populations, that is an allele is at high frequency because the individuals carrying it are favoured in both populations, then the allozyme data will be interpreted as indicating that the populations are not genetically different. On the other hand, two populations with habitat differences but which regularly exchange genes via larval flow may have different allele frequencies at a locus which is under selection where one allele is favoured in one population but is deleterious in the other. Differential mortalities in the two populations will cause allele frequency differences between samples taken from the two populations and this will give a false impression about the degree of reproductive separation of the two. This is why the use of several markers and, even better, different types of markers is recommended.

Here is a good example of how different makers can deliver a different message. The American oyster, *Crassostrea virginica,* occurs in estuarine habitats throughout the east coast of the USA, from Maine in the north to the Gulf of Mexico in the south. It has a larval phase lasting several weeks and might be expected to be relatively genetically homogeneous as a result of this. Allozyme studies detected no significant genetic differentiation into local populations anywhere in the species' range. However, mtDNA markers and some neutral, non-coding nuclear DNA markers demonstrated a very distinct division into a southern group and a northern group of populations separated at a zone of demarcation near to Cape Canaveral on the Florida Atlantic coast. This evidence makes it clear that there is practically no gene flow between the southern and northern regions, probably owing to the characteristics of a major oceanic current – the Gulf Stream – which emerges from the Gulf of Mexico around the tip of Florida. On the basis that evolution is faster in the mtDNA molecule than at protein coding loci, it could be suggested that, although differences in the mtDNA molecule have evolved, insufficient time has elapsed since the divergence of the two groups of oysters for allozymes to reflect the division. However, the non-allozymic single copy nuclear DNA markers also show this divergence, and the most convincing conclusion is that selection at the studied allozyme loci is maintaining the same alleles at similar frequencies on both sides of the break point. It is notable that several other species such as the horseshoe crab (*Limulus polyphenus*), the black sea bass (*Centropristis striata*) and the seaside sparrow (*Ammodramus maritimus*) also exhibit a fundamental mtDNA divergence in the same region, and this is congruent with the biogeography of the region.

What then are we saying here? Is it really the case that allozyme data, and more generally coding markers, should not be trusted at all? This is a vexed question, but the answer is that ideally, a large number of loci of different types should be used to provide a reliable picture of gene flow and population differentiation. Fortunately, selectively neutral markers such as mtDNA, microsatellites and SNPs, generally more rapidly evolving than allozymes, are now the tools of choice for studies of population differentiation.

mtDNA variation

An example of the use of mtDNA variation to detect population structure. The anadromous fish *Coregonus artedii*, the cisco, is important in the local fisheries of the river systems of the James and Hudson Bays in Canada. Allozymes were just not variable enough to be able to reveal differences between local river populations; therefore, a mtDNA approach was used by Bernatchez and Dodson (1990) to investigate population structure. The frequencies of three mtDNA clonal groups (A, B and C) of cisco from rivers feeding into the Hudson and James Bays.

The clonal groups are not individual mtDNA haplotypes, but come from three clear clusters on a dendrogram of the 19 haplotypes detected from the 141 fish sampled in the study. Different mtDNA haplotypes can be compared to estimate their relatedness to one another and this is what provides clusters of more closely related haplotypes. The genetic pattern revealed by a detailed mtDNA analysis of the different haplotypes within the clonal groups suggests that the cisco of this region, which was covered by ice in the last glaciation, were actually derived from two quite separate glacial refugia (clonal groups A and B) and have recolonised the area by different routes.

Since recolonisation, there has been some merging between the two types but this has not obscured the historical geographic pattern. The source of clonal group C which is strongly diverged from both the A and B clonal groups is not clear – it might even represent introgressed mtDNA from a closely related hybridising species. In this example, we see how mtDNA analysis can not only tell us about the current population structure of a species, but also can point to historical processes which have influenced the pattern observed in the present day. In human genetics, this valuable characteristic of mtDNA has been used to provide evidence for the 'out of Africa' theory, since it appears that we are all descended from a single 'mitochondrial Eve' lineage that inhabited Africa some 140 000–290 000 years ago.

Microsatellite variation

The rate of change at microsatellite loci is much faster than for allozyme loci or for mtDNA, and therefore, microsatellite variation is a more sensitive measure of weak genetic differences between populations. How much genetic differentiation can be resolved in salmon (*Salmo salar*) populations by microsatellites than with allozymes. The top dendrogram is based on an allele frequency analysis of 38 allozyme loci in populations from 31 rivers that drain into three main regions – the west Atlantic, the east Atlantic and the Baltic Sea. In the lower dendrogram, equivalent results based on four microsatellite loci are displayed (McConnell et al., 1995). The microsatellite genetic distances between river populations in the western Atlantic (D ranges from 0.06 to 0.11) are actually greater than the allozyme genetic distances when comparing salmon populations from both sides of the Atlantic Ocean (D = 0.04). So the power of microsatellites to identify population differentiation is very high. However, there are two caveats that must be added here. Firstly, the resolving power of loci with many alleles, such as microsatellites, is highly dependent on sample size. A small sample of, say, ten individuals scored at a locus with 25 alleles will inevitably only be sampling a proportion of those alleles. Comparing this with another small sample from a different population that also has the same alleles at similar frequencies could show quite large genetic differences based on chance alone. Some alleles sampled from one population will be absent in the sample from the other population, leading to the mistaken conclusion that the two populations are very different. Secondly, Nei's genetic distance measure *(D)* was devised on the basis of estimated point mutation rates at allozyme loci (around 10^{-7} mutations per gene per individual). As mutation in microsatellites is due to slippage during replication (around 10^{-3} mutations per gene per individual) rather than point mutation, it would not be safe to use Nei's formula (time of divergence = $5 \times 10^6 D$) to extrapolate any times of divergence between taxa based on microsatellite data.

POPULATION STRUCTURE IN THE FLAT OYSTER

As a final example of the use of genetic markers to investigate population differentiation we will look at the European flat oyster. The fishery for this oyster, *Ostrea edulis,* has a history stretching back over 2000 years. Indeed, the Romans imported these oysters from Britain and cultured them in lagoons in Italy (in 50 bc Sallust wrote 'Poor Britons – there is some good in them after all – they produce an oyster'!). Because of the high abundance of oysters in littoral and shallow sublittoral habitats and the ease of their management and capture, oyster fisheries have been an important local maritime activity in many parts of Europe for centuries. The arrival of the railways in the nineteenth century enabled the rapid transfer of oysters to the centres of industry and this species then became a very significant source of protein and a highly valuable business activity for European populations.

By the early mid-twentieth century, however, various factors such as overfishing, disease and habitat degradation had caused a massive collapse of most of the fished flat oyster populations. The 'Portuguese' oyster had been introduced from Asia to Portugal (probably in the sixteenth or seventeenth century) and then accidentally introduced to France in 1868, while the Pacific oyster was introduced in the 1960s and 1970s. These two species now supply much of the oyster market. Flat oyster populations, however, have never recovered to anything approaching their previous abundance.

So, the flat oyster has undergone two processes that could influence its population genetics: centuries of unrecorded human translocations of adults and seed together with more recent population bottlenecks. Leaving aside anthropogenic influences, we would expect a certain degree of genetic separation between populations that would relate to the dispersal potential of the larval stage (H"10 days) and the geographic separation between populations. This is the isolation by distance concept where increasing genetic distance is expected to correlate with increasing geographic distance. Of course, naturally occurring isolation by distance can be disrupted by human translocations, so it might be instructive for us to look at the population structure of this species.

The spatial genetic structure of the European flat oyster across its range from the Black Sea to Scandinavia has been investigated using three types of genetic marker: allozymes, microsatellites and mtDNA.

Allozymes

Nineteen European populations throughout Europe were scored at 14 polymorphic allozyme loci. Some of the loci showed patterns of variation over geographical distance while others did not. Three situations were seen: at some loci there was a single cline, showing a decrease of the frequency of the most common allele in north-western populations with distance south-east; one or two loci showed a 'V'-shaped pattern of frequencies, with opposing clines either side of the Strait of Gibraltar; many of the loci showed no strongly defined or consistent pattern of change in allele frequencies over geographical distance.

The simple uni-directional clines could reflect a selective response to environmental gradients or they could be the result of secondary contact between previously isolated populations that have different allele frequencies due to genetic drift. If the latter were the case, then the mid point of the cline should be at the position of the barrier which separated original stocks. In this case, the likely barrier would be at the Atlantic entrance to the Mediterranean, so this explanation does fit the loci that show a simple cline. However, the V-shaped cline is not easily explained by this idea of secondary contact between two different stocks and this suggests that selection could be operating at loci showing this pattern.

Although individual allozyme loci can show different patterns of spatial variation, an average G_{ST} of 0.088 and an average Nei's genetic distance (D) of 0.017 across all pairwise comparisons does indicate that there is some genetic heterogeneity. Whatever might have been the historical stock structure, the allozyme data resolve three main groups of populations: a North Atlantic group, a group that spans the South Atlantic and the western Mediterranean and finally an eastern Mediterranean group. Allozymes show a clear positive relationship between genetic and geographic distance although the slope of the graph is much shallower if the two allozyme loci showing strong clines (ARK and AP-2) are removed from the data set.

Microsatellites

Fifteen of the populations that had been studied with allozymes were scored for variation at five microsatellite loci.

The level of genetic differentiation estimated by F_{ST} (= 0.019) was much less than the equivalent G_{ST} (= 0.088) for allozymes. Also other microsatellite studies of more localised northern populations show a higher F_{ST} (0.05) so there may be some substructure that is missed by the wider scale sampling. Microsatellite data show a much-reduced correlation between genetic distance and geographic distance, but still a significant positive relationship that supports the idea of isolation by distance There is no evidence from the microsatellite data to support the idea of there being originally two separated stocks that have undergone secondary contact – simple isolation by distance seems more likely. As we saw with the American oyster, selection at allozyme loci can lead to incorrect conclusions based on indices of genetic differentiation because such indices were developed on the assumption that the loci are neutral.

An interesting feature of the data is that both allozymes and microsatellites exhibit generally lower genetic variation in Atlantic populations compared with those in the Mediterranean. It is possible that this is a result of smaller effective population sizes in the Atlantic caused by a shorter season reducing reproductive output, greater fishing pressure or the higher prevalence and destructive effect of an important parasitic disease, *Bonamia ostreae*.

DNA SEQUENCE VARIATION

The crude extraction of DNA from animal or plant tissue is a rather simple process, which involves mechanically or chemically breaking down the insoluble cellular structures and removing them by centrifugation. Soluble cellular proteins, and the proteins which bind the DNA into the chromosomes, can then be broken down using a strong protease enzyme and removed, usually using solvents such as phenol-chloroform. The DNA is present in the water-soluble component and can then be precipitated using an alcohol.

There are a number of commercial kits on the market which enable further purification of DNA. The next problem is to produce multiple copies of specific fragments of DNA, and this can be done either by cloning the fragment or by the use of the PCR. In the process of cloning, the target DNA is inserted into a vector molecule which is taken up or inserted into host cells. Subsequent rapid replication of these host cells and the vector molecules inside them results in the production of millions of copies of the target DNA. As far as most DNA markers are concerned, cloning is usually only needed during the development phase – once the DNA sequences flanking the markers have been found from the cloned fragments, PCR can be used to produce millions of copies of the target sequence within a few hours.

The PCR method relies on the fact that double-stranded DNA becomes denatured and separates into single strands when heated above 90°C. Once denatured, the temperature is lowered to a predetermined annealing temperature which allows short manufactured lengths of single-stranded DNA of known sequence (primers), designed to be complementary to the regions flanking the target DNA, to attach (anneal) to these flanking regions. Raising the temperature to 72°C in the presence of a DNA polymerase enzyme and the building blocks of DNA results in two copies of the double-stranded target DNA. Each time the cycle is repeated the number of copies is doubled, and since each cycle takes only a minute or two, millions of copies can be produced within a few hours by this method.

Let us first consider a mutation within the coding part (exon) of a gene that codes for an enzyme. We might expect such DNA mutations to have important effects. However, the mutation could occur at the third base of a codon and, because of the redundancy of the genetic code, will be unlikely to change the amino acid coded for. Non-synonymous mutations change the coded amino acid while synonymous mutations do not. Even though non-synonymous mutations change one of the amino acids in the enzyme coded for, this may not have any effect on the ability of the enzyme to carry out its cellular biochemical function. Nevertheless, *some* mutations within the exon of an enzyme gene are bound to have a deleterious effect such that individuals carrying that mutation produce an ineffective enzyme and are less likely to survive. Exceptionally, a mutation might be advantageous and improve performance of an enzyme. So enzyme exon DNA sequences are free to change slowly over evolutionary time, at a rate that is considerably less than the rate of mutation, and the rate varies between different enzymes depending partly on the specificity of their biochemical task in the cell.

What about DNA sequences which form part of an intron? These sequences are not translated into a protein product, and so we would expect changes to have neither deleterious nor advantageous effects. Mutations at non-coding sites are effectively neutral and therefore are likely to accumulate without constraint over evolutionary time.

Finally, let us consider sequences that code not for proteins, but for the very RNA molecules which are involved in the process of translation of the DNA code. Here, almost every letter of the code is critical to the functioning of the RNA product, and almost any mutation will render it non-functional. The strongly deleterious effect on any individual subjected to such a mutation means that the observed rate of evolutionary change of these parts of the DNA molecule is extremely slow. Such DNA is said to be highly conserved because most mutations are quickly rejected by natural selection.

It follows from the three examples above that some regions of DNA are valuable for identifying evolutionary changes far back in time, while others will detect more recent changes. Variations at a single base position in DNA are called single nucleotide polymorphisms and this term usually includes 'indels' (where a base pair is inserted or deleted). SNPs are generally bi-allelic – that is they usually have only two variants – and this restricts their use as genetic markers in some respects. However, recent technologies involving the production and screening of DNA micro-arrays have enabled easy identification of hundreds of thousands of single base pair variants throughout genomes. This has elevated SNPs to a high status as genetic markers for producing high-density genome maps, and they are increasingly being used in aquaculture species.

DNA FRAGMENT SIZE VARIATION

At the beginning of this chapter, we said that genetic variation can be measured and quantified at several levels. We have shown how we can determine the precise sequence of a length of DNA, and how it varies between individuals. Now we shall progress to see how differences between sizes of

DNA fragments can be identified and used to address particular genetic questions. Techniques that fall into this category include those known by the acronyms RFLP, VNTR, DNA fingerprinting, RAPD and AFLP. Of these, VNTR markers (microsatellites in particular) have come to the fore in recent years as being the most generally useful, though the others all have their place in answering particular genetic questions.

Restriction fragment length polymorphisms (RFLPs)

We can make good use of fragments of DNA as genetic markers without going through the procedure of sequencing them. If we have a high copy number of a particular fragment produced by the cloning method, from the PCR machine, or directly from mitochondrial DNA, this can be incubated with a number of different restriction endonucleases (REs) which will cleave it into a number of lengths depending on the position of the RE recognition sites. The various lengths produced can be separated by size and stained on an agarose gel using ethidium bromide (electrophoresis). The same piece of DNA from different individuals will produce different sets of restricted fragments if there have been point mutations (e.g. SNPs) affecting the RE recognition sequences. In this way, polymorphisms can be identified based on the pattern of the size fragments on the agarose gel.

Variable number tandem repeats (VNTR)

Variation in the sequence of DNA can occur at certain sites by a method which is not point mutation. Spread throughout the genome are regions called VNTR, which contain tandem (i.e. linked in chains) repeats of DNA sequences. The sequences may be very short (from 1 to 10 bp) or much longer, but the key feature of these tandem repeats is that the number of repeats can vary between individuals. It is thought that increases or decreases in the number of the repeats occur during copying due to interference during recombination events or due to replication slippage and that these processes are not only independent of point mutations but also occur at a much faster rates. Variation in the number of repeats at these satellite (repeated units 100–5000 bp), minisatellite (repeated units 5–100 bp) or microsatellite (repeated units 2–4 bp) loci can be very extensive in populations and provides a valuable tool for investigation of population genetic changes in the recent past. Microsatellite markers in particular are now used extensively for a number of reasons: because they are co-dominant (both alleles can be identified) and therefore can be analysed under the standard Hardy–Weinberg model; because, as they seldom occur within coding DNA, they can usually be considered to be free of selective pressures; because of the high number of both loci and alleles at each locus; and, not least, because automatic DNA sequencers can be used for automated genotyping at microsatellite loci, vastly increasing the rate at which samples can be processed.

DNA fingerprinting

DNA fingerprinting can be thought of as a combination of RFLP and VNTR. Firstly, genomic DNA is cut with a particular suite of restriction enzymes and differently sized fragments are separated by electrophoresis. The DNA is transferred from the fragile gel to a nylon membrane by the technique known as Southern blotting, and the membrane is then probed (hybridised) with a particular labelled satellite repeat sequence which is common throughout the genome. Fragments that contain the repeat show up as a number of discrete labelled bands. These banding patterns are so variable as to be in practice unique to each individual (the chances of a match between unrelated individuals are millions to one). Since the bands are inherited in a predictable fashion, DNA fingerprinting is a very accurate way of determining parentage and this method was one of the first ways in which DNA data were used in police forensics.

Random amplified polymorphic DNA (RAPD)

The RAPD method is based on the principle that the shorter the length of the primers which are used in PCR, the greater is the chance that non-target sequences will be amplified. Using a single 10-mer oligonucleotide as the sole primer, PCR is conducted on raw DNA and the resulting fragments, which come from annealing of the primers all across the genome, are separated on agarose gel and stained with ethidium bromide.

Variations between individuals in the presence or absence of bands reflect mutational differences (e.g. SNPs) at the primer sites. RAPDs suffer from the important criticism (among others) that they are neither entirely reliable nor repeatable.

Amplified fragment length polymorphism (AFLP)

A method which amplifies randomly selected fragments of the genome much more reliably than RAPD is the AFLP method. AFLP is almost an inverse form of RFLP – the genomic DNA is cut into fragments with restriction enzymes and then just a few of those fragments are selectively amplified using special labelled PCR primers. The resulting fragments are then visualised after separation on a polyacrylamide gel.

The products of AFLP – a series of bands of different sizes – are similar to the products of RAPDs, and variation between individuals is based on the presence or absence of bands. Variation is the result of point mutation differences in the DNA sequence at the PCR primer sites.

PROTEIN VARIATION

So far we have considered genetic variation at the level of the DNA. However, DNA sequence variation, when transcribed, can give rise to differences in the resulting proteins. It is at this level that genetic variation begins to interact more directly with the environment to affect the survivorship and reproduction of organisms and their genes.

Genetic variation at the level of proteins can be identified and quantified using electrophoresis to separate the different protein products of alleles followed by staining to visualise these protein products. It is possible to stain the gel to display all proteins, but it is more useful to take advantage of the substrate-specific catalytic abilities of the class of proteins known as enzymes.

This involves using the specific substrate of an enzyme in a stain overlaid on the gel that will change colour where the substrate is altered by the enzyme. The position of any enzyme variants can therefore be located on the gel.

These genetic variants of enzymes are known as allozymes, and methods for detecting allozyme variation were first developed in the 1960s. Although nowadays regarded by some as an outdated method, the extensive allozyme data sets produced in the last 40 years of the twentieth century fundamentally shifted the ground upon which geneticists tread.

From the practical point of view, allozymes enabled us to look at the genetics of natural populations of a whole range of aquatic organisms in a way that was never possible before. (The influence of allozyme data will become evident in several chapters of this book.) This is not to say that allozymes are only of historical interest – they are still a useful tool in answering many genetic questions, particularly given the infrastructure investment required to utilise the current sometimes bewildering variety of modern DNA technologies.

Although little used in fisheries or aquaculture research, the reader should be made aware of the technique of immunological testing which assesses the relationship between proteins on the basis of the relative strength of the antigen–antibody reaction that they will produce.

Phenotypic variation

There are very few examples of easily identifiable phenotypic variation in aquatic organisms controlled by single genes, or even pairs of genes. The best examples are found in the colouring of ornamental fish. It is interesting to consider just how lucky Gregor Mendel (the nineteenth-century discoverer of the method of genetic inheritance) was to have chosen to work with peas, which had a number of easily identifiable characters (round or wrinkled seeds; tall or short plants) each controlled by single genes. Such easily identifiable single-gene phenotypes are rare in most organisms. Because of the extensive development of protein and DNA genetic markers since the 1960s, the search for single-gene phenotypic variation is now uncommon. However, it is important to realise that the visual identification of varieties can be of critical importance to fish farmers without access to a modern genetic laboratory. Phenotypic rarities can provide high-value niche markets, and understanding how to get them to breed true requires genetic knowledge.

LEVELS OF GENETIC DIFFERENTIATION IN AQUATIC ORGANISMS

Many studies have been carried out to estimate the amount of genetic subdivision within aquatic species. Levels of population substructure in various groups. On average, these levels of G_{ST} (which is equivalent to F_{ST}) are not too far from 0.2, suggesting that, in general, subdivision of populations within species is close to that which allows genetic drift to change allele frequencies in local populations in the face of gene flow by migration. However, these average values hide a wide range of variation in population structure between species. For example, let us take two contrasting cases – the salmon, *Salmo salar*, and the mussel, *Mytilus edulis*. The salmon has a life history involving extensive juvenile and adult migration, with mature adults returning to their natal rivers to spawn. Thus, although salmon can be found over a very wide area, they are genetically very much restricted to local populations in their own rivers as this is where the transmission of genes between generations occurs. There are strong barriers to gene flow because very few fish return by mistake to non-natal rivers. In addition, each local population consists of relatively few individuals and the population can be sometimes even further subdivided into year classes that return to spawn in the same year as one another. Values of F_{ST} for salmon, calculated from extensive allozyme data, are in the region of 0.4 and indicate that, as expected, there is very little gene flow ($N_e m$ = 0.38) and strong genetic differentiation into local populations in this species. On the other hand, the mussel is a sedentary organism and adults never actively move more than a very short distance. As with salmon, mussels release eggs and spermatozoa into the water, but, in contrast to salmon, mussel larvae are small and planktonic and are at the mercy of currents that can disperse them over great distances. Larval life, from egg to metamorphosis, in mussels lasts around 4 weeks and can be further extended in two ways. Firstly, if a suitable habitat for settlement is not discovered, metamorphosis can be delayed for more than a week, allowing further dispersal. Secondly, once metamorphosed, these very young mussels (spat) can detach, secrete a very thin byssal thread and be transported in currents by a process known as byssal drifting, which is analogous to the aerial dispersal of spiders by gossamer threads. Typically, F_{ST} values from allozyme studies in mussels are very low (<0.001) and clearly reflect extensive gene flow ($N_e m$ > 250) and a lack of any population substructure.

It is also interesting to compare genetic differentiation in marine pelagic fish such as Atlantic herring (*Clupeus harengus*) with that in anadromous fish such as salmon. The herring has a very large effective population size because all fish return to the same region to spawn and there are therefore few, if any, barriers to gene flow. Average F_{ST} for herring is 0.01 and $N_e m$ is 24.8.

Therefore, the conclusion is that where the effective population size and/or migration is large, gene flow will tend to dominate over random genetic drift, there will be little differentiation and F_{ST} will be close to 0. Where $N_e m$ is small, random genetic drift will tend to dominate over gene flow, allele frequencies will differ strongly between populations and F_{ST} will be large.

Although information about population differentiation is very valuable to fishery managers, it is important to recognise that a *lack of evidence* for local genetic populations does not always mean a *lack of substructure*. If populations are genuinely genetically isolated from one another, then we would expect random genetic drift to be acting to change allele frequencies at all polymorphic loci. However, for reasons of chance, not all loci will show different allele frequencies in different populations. If the loci screened in limited surveys happen to be these invariant loci, then the true variation remains hidden. It is therefore important to examine sufficient loci to be sure that an apparent lack of genetic differentiation can be relied upon.

MIXED STOCK ANALYSIS (MSA)

When an oceanic salmonid catch is brought ashore, it probably consists of fish from a mixture of distinct populations from different river sources. We have already seen how salmonids can be genetically very divergent between rivers and these differences are likely to be reflected in factors such as growth rates, age at maturity or other life history parameters. If the fishery is managed as a single unit, there is the potential for the overexploitation of some populations, or the suboptimal harvesting of those that are more abundant. In order to enable proper management and obtain the maximum sustainable yield from such fisheries, we need to estimate the proportions of the various populations in the fished resource. If the contribution of the various populations varies geographically or over time, then fishing effort can be directed to exploit the strongest populations, while pressure on weaker populations can be reduced, by targeting fishing to particular regions or particular times.

Early attempts to determine the relative contributions of individual populations to a fishing stock by tagging or branding fish as they departed from specific rivers were relatively ineffective because of the high initial labour costs and the loss of markers in a proportion of the fish.

Far more effective is the use of genetic markers. Allozyme markers have been widely employed in the management of Pacific salmon, using comparisons between allele frequency data for the individual river populations and allele frequencies in the total catch.

Of course, to make it work, the differences between river populations must be sufficiently great that allele frequencies at minimally a few of the loci scored will enable identification of that population in the total catch. Fortunately, there are statistical methods available which made these comparisons possible even when allozyme allele frequency differences are not very great. However, the potential resolving power of microsatellite markers is much greater than allozymes, and they are now the markers of choice for all studies of MSA.

MSA is also relevant to some sport fisheries that consist of a mixture of natural and introduced stock. The introduced element is either transplanted from another area or derived from hatchery production. A good example is the trout, *Salmo trutta*, which exists in Europe in two different forms. One, the resident or brown trout, inhabits rivers and lakes and never goes to sea, while the other, the sea trout, is anadromous and, as its name suggests, migrates into the sea and spends some time there before returning to its natal river to spawn. Where waterfalls that are impassable to trout are present in rivers, the populations above the falls will be resident trout and those below will be sea trout. Because sea trout grow bigger than resident trout, they make a better fighting fish for rod and line and, mainly for this reason, restocking of river populations above impassable falls is often carried out with hatchery-reared sea trout.

The expectation was that there would not be a problem with the survivors of these introduced sea trout breeding with the natural population of resident trout because, once they had migrated to the sea, they would not be able to return to their original position (above impassable falls) in the river to breed. However, allozyme and microsatellite-based studies in various parts of Europe have shown that this does not always happen. It is clear that some stocked sea trout do not migrate to sea but remain in the populations of resident trout with which they will then breed. This leads to a loss of the unique genetic identity of the resident stocks and is an important conservation issue.

In North America, the cutthroat trout exists as a complex of up to 15 different recognised subspecies, and there has been extensive introduction of the different subspecies into non-native habitats. Genetic marker analysis has demonstrated that this has resulted in an extremely complex pattern of hybridisation and introgression. This can be bad news because the hybrids often suffer developmental abnormalities that make them less fit than the pure-bred subspecies. Furthermore, as in the case of brown trout in Europe, there is evidence of genetic swamping which threatens the uniqueness of locally adapted populations.

Stocking of rivers for salmon fishing using non-indigenous fish has been extensively carried out across the world. Some rivers are thought to contain no descendents of the original indigenous populations.

Such eradication of the native population was believed to be the case in many Danish rivers until analysis of microsatellite data from archived scales was used to identify indigenous salmon.

DNA was extracted from adipose fin samples from recently caught fish from various Danish rivers and from old scale samples (archived between 1913 and 1954) from some of these rivers. Samples were scored at six microsatellite loci and analysed using a 'self-classification test' that enabled individual fish to be 'assigned' to particular populations based on their multi-locus microsatellite genotypes. Information about the microsatellite alleles and their frequencies in indigenous river populations were obtained from the scale samples which pre-dated the extensive introductions of non-indigenous salmon. In spite of these extensive introductions, results revealed that around 11% of fish currently in the rivers had genotypes that were characteristic of the indigenous fish. This showed that descendents of indigenous fish were still present and identifiable and had not become genetically altered or homogenised by extensive hybridisation with exogenous fish.

Using non-invasive DNA sampling of salmon (small fin clips or blood samples that do not harm the fish) to identify indigenous individuals, it has been possible to undertake breeding programmes for re-stocking rivers with native stocks. This study demonstrates the power of microsatellite variation as a tool for MSA that can identify indigenous fish and allow restoration of locally adapted stocks. Now we will consider how population size can impact on the genetics of organisms. The numbers of individuals within a population that contribute to the next generation is a key factor in the maintenance of genetic variation, and there are two situations where this will be of relevance. Firstly, small population sizes can occur in the wild due to natural physical (e.g. climate extremes) or biological (disease) forces and also due to anthropogenic effects such as pollution or overexploitation. Secondly, small population sizes are a necessary element of aquaculture that involves the use of a hatchery to manage the production of young fish and shellfish. Therefore, the genetic concepts and constraints focused on the conservation of scarce species, or small endangered populations within a species, are very similar to those that are relevant to hatchery production in aquaculture.

The concept of random genetic drift, that is the process that causes random changes in allele frequency from generation to generation. When the effective population size (N_e) is very large, changes in allele frequencies between successive generations will be very small. On the other hand, in populations that have a small effective size, there will be a large variance between generations in the frequencies of alleles, that is natural fluctuations in allele frequency between generations will be much greater in small populations. In such populations with small effective size, this will lead to a reduction in genetic diversity over time that can be identified as a loss of alleles and loss of heterozygosity. Why does it matter that alleles might be lost? Each variant allele at each coding locus in a population can be regarded as part of the 'genetic resource' of that population. An allele alone, or in combination with other alleles or loci, could be responsible for conferring on its carrier a valuable trait such as increased resistance to a particular disease, better cold tolerance or faster growth.

Therefore, the loss of any allelic variants is a potential loss of valuable genetic resource. Of course, if most allelic variation at coding loci is neutral, then this is less important, but we would be unwise to ignore the certainty that at least *some* variants at coding loci will be advantageous. If not now, then most likely in the near- to medium-term future, global warming will bring about the increasing importance of high-temperature-resistant allelic variants at biochemically important loci in temperate aquaculture species. Such alleles may be effectively neutral until extreme summer temperatures reveal their value. A second important consequence of small effective population size is the phenomenon of inbreeding that is brought about through matings between closely related individuals. From the genetic perspective, inbreeding increases homozygosity and almost always has deleterious phenotypic effects, making inbred offspring less likely to survive than non-inbred offspring. The phenotypic consequences of inbreeding (low viability, poor growth, abnormalities) are labelled inbreeding depression.

Mitochondrial DNA

The method of single-strand conformational polymorphism was used to identify mtDNA haplotypes from the same 15 populations of European flat oyster as had been used for allozyme and microsatellite analysis. Single-strand conformational polymorphism involves denaturing polymerase chain reaction (PCR)-amplified DNA (in this case a 313-bp fragment of the 12S-rRNA gene) at 95æ%C before running the product down an acrylamide gel.

The final position on the gel of the single strand DNA is identified by ethidium bromide.

A total of 14 haplotypes were identified. One haplotype predominated in the Mediterranean, while another prevailed in the Atlantic populations. A third haplotype was common in samples taken at the geographic extremes of the species distribution. So the mtDNA analysis generally supports the idea of an Atlantic stock and a Mediterranean stock of flat oysters.

However, unexpectedly, some genetic similarity is shown between Norwegian and Black Sea oysters and this is difficult to explain. The overall pattern of isolation by distance is much more strongly demonstrated by the mtDNA data than with allozymes and microsatellites. The average F_{ST} is 0.244, an order of magnitude greater than that for microsatellites.

How can this difference be explained? For most animals, because mtDNA is haploid and maternally inherited, there is a (theoretically fourfold) smaller effective population size compared with diploid nuclear markers. This means that mtDNA variation is likely to be four times more sensitive than nuclear markers to factors that reduce population size and create bottlenecks (such as overfishing, habitat loss or disease infestations). However, in flat oysters, this would be reduced to twofold because they are sequential hermaphrodites and individuals can be both male and female in their reproductive lifetime.

There are two other factors that might possibly contribute: unbalanced sex ratio and/or variance in female reproductive success. Flat oysters usually develop first as males and only in later years change to produce female gametes.

Within any particular season there is generally a 3:1 male to female ratio. This deviation from the 1:1 male to female ratio is probably exacerbated in areas where there is a high prevalence of the parasite *Bonamia ostreae,* because highest mortalities from this parasite usually occur in 2–3-year-old adults.

These older oysters are more likely to be female than male, which increases the proportion of males in the genetically effective population. An unbalanced sex ratio will reduce effective population size. Variance in individual reproductive success is large in organisms, like oysters, that produce huge numbers of eggs and such variance is thought to be greater in females than in males.

This variance between individuals could be an important component in reducing effective population sizes of oysters. So there are some reasonable biological explanations for the magnitude of difference in estimations of population differentiation based on mtDNA and microsatellite data.

Populations at the extremes of the geographical range of a species are likely to experience stronger environmental pressures and greater temporal variability in reproductive success than populations well within the species range. This is expected to result in an increased likelihood of bottleneck events and overall smaller effective population sizes.

We can see the genetic consequences of this in these studies in oysters: at allozyme loci, microsatellite loci and in mtDNA markers, the most northern population from Norway and the most eastern population from the Black Sea both have generally less genetic variation (fewer alleles and lower heterozygosity, fewer haplotypes) than other populations.

Overall then, we can see how different genetic markers can throw up different estimations of population differentiation and that the analysis and interpretation of data from these different markers must be carried out with reference to the biology of the organism.

GENETICS OF SMALL POPULATION SIZE IN THE WILD

Most educated people are now aware of the fragile state of the planet and the increasing pressures from human activities on the species with which we share the biosphere.

Species are becoming extinct at a rate comparable to the mass extinctions of geological time, and in addition to loss of individual species, where population sizes have fallen due to human influence (loss of habitat, overexploitation), there is a loss of genetic biodiversity within remaining species. This loss of genetic variation within species can be identified as loss of alleles and a reduction in heterozygosity.

One of the difficulties in assessing genetic diversity in wild populations is that some marine species exhibit the phenomenon of chaotic patchiness. This is the situation where there is extensive microspatial variation in allele frequencies detected at any one sampling time, but allele frequency changes occur over time such that the pattern observed might be very different if sampled at another time.

Chaotic genetic patchiness is more common in the marine environment than others because of the explosive reproductive capacity of just a few individuals, the uncertainties and vaguaries of larval dispersal, and the mosaic nature of marine and littoral habitats. Therefore, the lucky survivors of spawning, larval dispersal and final settlement are seldom the average genetic representatives of the parent population.

This has been called the 'sweepstake concept': the reproductive success of the minority and the reproductive failure of the majority, and was first demonstrated in Pacific oyster (*Crassostrea gigas*) populations and then observed in many other species. What this means is that for particular populations, or across the species as a whole, the effective population size is actually much smaller than the census population size. Until recently it was thought that heavily overexploited fish stocks would still be sufficiently abundant to avoid the sweepstake effect and overcome the danger of decreased genetic diversity. However, several studies have now identified unexpectedly low effective population sizes (N_e) in commercial marine fish species that normally have very large census population sizes.

One approach used to estimate the effective population size of fish has been based on extracting DNA from modern fish samples and from archived otoliths or scales of various ages and comparing genetic diversity and heterozygosity at microsatellite loci over time.

When the effective population size is large, there is a balance between the loss of microsatellite alleles through genetic drift and the introduction of alleles by replication slippage. If the effective population size falls below a certain level, the balance shifts so that the loss of alleles through genetic drift is greater than the rate of introduction of alleles.

Loss of allele over time also leads to loss of heterozygosity. Using mathematical models based on temporal fluctuations in allele frequency and decrease in heterozygosity, it is possible to estimate the effective population size of a population.

The first example we shall look at is the New Zealand snapper *(Pagrus auratus)*. DNA was extracted from snappers fished in 1998 and from dried snapper scales that had been collected and archived between 1950 and 1986. The DNA was genotyped at seven microsatellite loci, and the effective population size was estimated from temporal changes in allele frequencies and from the decrease in heterozygosity over time .

The census size estimation for this fish stock ($N = \sim 6.8 \times 10^6$) is five orders of magnitude higher than the estimated effective population size (somewhere between 80 and 770), giving an $N_e{:}N$ ratio of approximately 2.6×10^{-5}.

Our second example is based on microsatellite analysis in the plaice, *Pleuronectes platessa.* Using archived otoliths from 1924 to 1972 and samples from juveniles caught in 2002, the effective population size for plaice was estimated to be 20 000 in the North Sea and as low as 2000 in Iceland.

As with the New Zealand snapper the $N_e{:}N$ ratio approached 2×10^{-5}. Samples of plaice from before the 1960s, as well as being more genetically variable, were generally in agreement with the Hardy-Weinberg model, while more recent samples usually exhibited a significant deficiency of heterozygotes.

The heterozygote deficiencies are believed to be a signal of potential inbreeding consequent on low effective population size . Identification of an effect that appears to link back to the steep increase in fishing pressure since the 1960s shows how overfishing can have genetic consequences long before stocks become identified as 'at risk'. The third example is from a localised North Sea cod fishery off Flamborough Head on the north-east coast of the UK.

Cod, *Gadus morhua,* has been a hugely significant commercial fish in Europe and North America for centuries, but now, because of massive reductions in stock sizes, this important fish is on the *IUCN Red List of Threatened Species.* This study investigated microsatellite genetic variation in archived otoliths from fish born in 1954,1960,1970 and 1981, together with live fish caught in 2000 (born in 1998).

Earlier samples between 1954 and 1970 show a loss of allelic diversity over time, but later samples show a reversal of this trend. The earlier samples are consistent with genetic drift in a population with reduced effective population size, estimated to be an average of around 100 (giving an N_e:N ratio of 3.9 x 10-^{5}) between 1954 and 1970.

However, the increase in microsatellite allele diversity after 1970 included alleles not originally present in samples prior to 1970, which suggested that there had been an influx of fish from neighbouring regions. In fact, by 1998, the Flamborough Head cod consisted predominantly of immigrants from neighbouring populations. Although genetic diversity seems to have been restored, local adaptations may have been lost or compromised.

These examples demonstrate that the effective population size of commercially exploited marine fish populations is unexpectedly low relative to the total number of fish and that N_e:N ratios are in the region of 2 x 10-^{5} to 4 x 10-^{5}.

This low effective population size has led to a reduction in genetic diversity and a higher risk of inbreeding depression. These factors reduce the ability of populations to recover from overfishing bottlenecks and limit their ability to respond to environmental change.

Because of the life history of anadromous salmonids that sees them returning to their natal rivers to spawn, these species are divided into many small breeding groups or populations with limited gene exchange. They are therefore more likely than other marine fish to suffer from small effective population sizes.

A recent study has used microsatellite genetic variation and efficient computer analysis programmes to identify the kin relationships between individuals in two endangered wild populations of salmon (*Salmo salar*) in Canada. One population came from the Stewiake River and the other from the Big Salmon River, and microsatellite markers show big differences between these two rivers.

Collections of fish from these rivers were made in 1998 to establish captive breeding and rearing programmes in an attempt to retain their distinctive genetic nature. Using nine microsatellite markers, individuals were genoytyped and these genotype data were analysed using modified specialist computer programmes for pedigree analysis. Without having any direct parental information the analyses placed individuals into half-sib kin groups that were further divided into full-sib families.

In the Stewiake River, 56 kin groups were identified, and most fish belonged to just a few of these kin groups. Indeed, nine of these groups had only one individual assigned to them. Similar results were obtained for fish from the Big Salmon River.

In each respective half-sib kin group the fish were generally distributed into several small full-sib families, and this is probably indicative of the well-documented mating behaviour in this species, where more than one adult male and several additional precocious small males can fertilise the eggs of a single female in her redd.

Knowing the precise number of families in the samples enabled a good estimate of the effective population size of the parental generation to be made. The estimated N_e values for the Stewiake River and the Big Salmon River were 68.8 and 73.4, respectively, which were much lower than the census population size. These highly informative data show that both unbalanced sex ratio and variance in maternal family size contribute strongly to the small effective population size relative to the census population size in this species.

Considering commercial marine fish and shellfish species, complete extinction directly as a consequence of overfishing is an unlikely outcome because fishing pressure is usually reduced when it becomes uneconomic to target a particular species.

While it would be comforting to think that there could still be sufficient numbers of individuals of overfished species out there to maintain small populations, the low effective population sizes identified in these studies emphasise the real genetic risks following overexploitation.

Of course, when numbers of individuals in a species fall to very low levels, the problems of loss of diversity, exacerbated by inbreeding depression, can lead to extinction. Artificial breeding programmes may then become options we show how breeding plans can be developed to maintain the greatest genetic diversity in spite of small population size.

The types of fish species, which are even more vulnerable to extinction from overfishing or habitat loss, are those which have colonised lakes following the retreat of the ice sheets of the last glaciation. Such species are often strongly substructured, with particular subspecies or variants restricted to just one or a few lakes. Finally, we must mention marine mammals which have been probably the highest profile group of threatened species in the marine environment. It is difficult to secure samples for genetic study from whales. Indeed, until the development of DNA techniques which required only minute quantities of tissue, such as can be obtained from the sloughed-off skin, little genetic information was known about cetaceans.

More recently, using mtDNA, microsatellites and DNA fingerprinting, together with observations of behaviour in the wild, much has been learned about the family relationships between individuals within pods. Such information has assisted the conservation effort, not least by revealing and publicising the complicated and subtle nature of the population substructure of these species.

Bibliography

Adams, C.E. : *Mammalian Egg Transfer*, Boca Raton, FL, CRC Press, 1982.

Alexander, R.: *The Chordates*, London: Cambridge University Press, 2001.

Allen, C.: *Cognitive Ethology and the Intentionality of Animal Behavior*, Bombay: Manaktalas, 2006.

Andrewartha, H.: *Introduction to the Study of Animal Populations*, London: Methuen Press, 2000.

Arti Sharma: *Fishes : Aid to Collection, Preservation and Identification*, Daya, Delhi, 2006.

Ausubel, F.M.: *Current Protocols in Molecular Biology*, New York, John Wiley and Sons, 1989.

Bains, William: *Biotechnology: From A to Z*, New York, Oxford University Press, 1998.

Barnes, R.: *Invertebrate Zoology*. Philadelphia: Saunders Co., 1980.

Bauer MW: *Biotechnology-the Making of a Global Controversy*, Cambridge, Cambridge University Press, 2002.

Bhatnagar, Vasudev : *Cell Science and Technology*, Campus Books, Delhi, 2009.

Brandwein, P.F.: *Sourcebook for the Biological Sciences*, San Diego, Harcourt Brace JOvanovich, 1986.

Broach, J.R. : *The Molecular Biology of the Yeast*, Cold Spring Harbor: Cold Spring Harbor Laboratory, 1981.

Bronson, F. H.: *Mammalian Reproductive Biology*, Univ. Chicago Pr., Chicago, 1990.

Burghardt, G.M.: *Animal Awareness: Perceptions and Historical Perspective*, Cambridge: Cambridge University Press, 2003.

Carroll, R. L.: *Vertebrate Paleontology and Evolution*. W. H. Freeman and Co., New York, 1988.

Case, V.: *Animal Behavior Class Presentation*, Delhi: Motilal Banarsidas Press, 2000.

Crawford A. : *Experiments and Observations on Animal Heat*, London: Printed for J. Johnson; 1788.

DeGrazia, David: *Animals Rights: A Very Short Introduction*. Oxford: Oxford University Press, 2002.

Devyani Khemka: *Animal Physiology*, Dominant, Delhi, 2003.

Dixon, Dougal: *After Man-A Zoology of the Future*, New York, St. Martin's Press, 1981.

Ebbesson, S.: *Comparative Neurology of the Telencephalon*, New York: Plenum Press, 2005.

Edward A.: *Papers on Bacterial Genetics*, Boston, Brown and Company, 1960.

Edward A.: *Papers on Bacterial Genetics*, Boston, Brown and Company, 1960.

Escobar, Roberto Calle: *Animal Breeding and Production of Camelids*, Lima, Peru, 1984.

Gardiner, M.S.: *The Biology of the Vertebrates*. New York: McGraw-Hill, 1980.

Goodsell, David S.: *Bionanotechnology: Lessons From Nature*, Hoboken, Wiley-Liss, 2004.

Gordon, G. A. : *Animals Physiology*, Harper and Row, New York, 1989.

Gould-Somero, M.: *"Echiura." in Giese and Pearse, Reproduction of Marine Invertebrates*, New York: Academic Press. 1975.

Gray, J.: *Animal Locomotion*, Norton, New York. 1968.

Griffin, D. R.: *Animal Mind-human Mind*, Berlin: Dahlem Konferenzen, 2001.

Hagedorn, A.L.: *Animal Breeding*, Crosby Lockwood, 1950.

Le Gros Clark: *The Fossil Evidence for Human Evolution*, Chicago: University of Chicago Press, 2001.

Marshall, A.J., and W.D. Williams: *Textbook of Zoology Invertebrates*, New York: American Elsevier Publishing Co., 1972.

Martin, A.M. : *Fisheries Processing : Biotechnological Applications*, Chapman and Hall, Delhi, 2009.

Montgomery, G. G.: *The Early Placental Mammal Radiation Using Bayesian Phylogenetics*, Science, December 2001.

Narvekar, Raghunath : *Molecular Biochemistry : Principles and Practices*, Adhyayan Pub, Delhi, 2008.

Old, R.W. : *Principles of Gene Manipulation*, London, Blackwell Scientific Publications, 1989.

Prasad, S.K. : *Biochemistry of Carbohydrates*, Discovery Pub, Delhi, 2010.

Primrose, S. B. : *Principles of Gene Manipulation*, London: Blackwell Scientific Publications, 1994.

Rathnakumar, K : *Fish Processing Technology and Product Development*, Narendra Pub, Delhi, 2008.

Romer, A.: *Vertebrate Paleontology*, Chicago: University of Chicago Press, 2001.

Roobeek A: *The Biotechnology Revolution?*, Oxford, Blackwell, 1995.

Seidel, S.M.: *New Technologies in Animal Breeding*, Orlando, FL, Academic Press, 1981.

Sharma, Anil : *Cell Biology, Genetics and Plant Breeding*, Indus Valley Pub, Delhi, 2006.

Sirks, M.: *The Evolution of Biology*, New York: Ronald Press, 2003.

Slater, R.J.: *Experiments in Molecular Biology*, Clifton, Humana Press, 1986.

Susan R.: *Biotechnology: An Introduction*, Belmont, Thomson/Brooks/Cole, 205.

Switzer, R.L.: *Experimental Biochemistry*, New York, W.H. Freeman and Company, 1977.

Teich, M.: *Fundamentals of Photonics*, New York: Wiley, 1991.

Towle, Albert: *Modern Biology*, Austin, Holt, Rinehart and Winston, 1988.

Tripathi, Nirmal : *Molecular Biotechnology*, Crescent Publishing Corporation, Delhi, 2008.

Walden, Richard: *Genetic Transformation in Plants*, England, Open University Press, 1988.

Winston, Mark L.: *Travels in the Genetically Modified Zone*, Cambridge, Harvard University, 2004.

Yablokov, A.: *Variability of Mammals: Moscow*, USSR, Nauka Publishers, 1966.

Index

❑❑❑